BANTAM User Guide

Springer

London
Berlin
Heidelberg
New York
Barcelona
Hong Kong
Milan
Paris
Singapore
Tokyo

Julian Ashbourn

BANTAM User Guide

Biometric and Token Technology
Application Modeling Language

 Springer

Julian Ashbourn

British Library Cataloguing in Publication Data
Ashbourn, Julian, 1952-
  BANTAM user guide : biometric and token technology
  application modeling language
  1. Object-oriented programming (Computer science) 2. Biometry
  - Computer programs
  I. Title
  005.1'17
ISBN 1852335130

Library of Congress Cataloging-in-Publication Data
Ashbourn, Julian, 1952–
  BANTAM user guide biometric and token technology application modeling
  language-BANTAM/Julian Ashbourn
    p. cm.
  Includes index.
  ISBN 1-85233-513-0 (alk. paper)
  1. Computer simulation. 2. Bantam. I. Title.
  QA76.9C65 A84 2002
  003'.3–dc21   2001042811

Additional material to this book can be downloaded from http://extras.springer.com.

ISBN 1-85233-513-0 Springer-Verlag London Berlin Heidelberg
a member of BertelsmannSpringer Science+Business Media GmbH
http://www.springer.co.uk

Typesetting: Electronic text files prepared by author
Printed and bound at the Athenæum Press Ltd., Gateshead, Tyne and Wear
34/3830-543210 Printed on acid-free paper SPIN 10838900

This book is dedicated to all those individuals who have struggled to implement new and emerging technologies as early adopters, often against a sea of ignorance or misinformation concerning device functionality, interfaces and other issues, and further hampered by poor communication and the lack of a common language with which to articulate their requirements.

Special thanks to Rosie, Karen, Sally, Melanie and all the team at Springer UK for their enduring support and encouragement throughout the development of the BANTAM User Guide.

Foreword

There has been a long standing joke that every year the business plan that goes to the board of directors contains a line item that states; 'This is the year for implementation of the biometric deployment plan.' It has long been a concern of mine that the same line will keep showing up in the business plan indefinitely and that the use of biometrics in day-to-day activities will occur around the time we have flying cars and colonies on some other planet.

Biometrics, long considered to be part of some futuristic fantasy or a page in a high security spy novel, has been around for a number of years. This being the case, I often wonder why we have not seen greater adoption of biometric authentication mechanisms in the new 'always connected' world we now live in. Where we aspire to provide the 'right information, to the right person, at the right time, anywhere, any device' the use of biometrics has become more needed now then ever before. As security and privacy issues move to the forefront of our daily lives, biometrics will move from the business plan to the deployment phase.

The official BANTAM user guide provides the information to bring the movement on use of biometrics to the next level. As the guide aptly states, there has been a lack of clarity as to what possibilities these technologies can bring. This guide goes beyond articulating the possibilities and addresses issues such as procurement, project management and training. Julian's book, with the accompanying CD-ROM and useful utilities, provides a methodology that the reader can easily use, and that will be of great value when dealing with biometric, token technology and related applications. I found the BANTAM user guide extremely informative and well written, and I am sure that you as readers will as well.

Howard A. Schmidt
Chief Security Officer, Microsoft Corporation
International Board President, Information Systems Security Association.

Preface

Biometric technology has been available in various guises for well over a decade. It has long since left behind the 'futuristic' image as depicted in spy or space movies, with many biometric techniques currently developed to advanced levels of both performance and usability. In addition, there have been many initiatives to integrate biometrics with synergistic technologies such as chip cards and public key infrastructures (PKI) among others. In view of this, one could be forgiven for wondering why we don't see more biometric applications in everyday life, especially given the predictions made for the technology in the early days. The reasons for this apparent anomaly may be many and varied, including general awareness, cultural perception, user psychology issues and others. One of the biggest stumbling blocks, however, has undoubtedly been the difficulty in precisely understanding and articulating requirements, and matching these to available products via published specifications and other available information: in short, the ability to understand the possibilities offered by biometric and related technologies, and design and implement applications accordingly. In addition, the available biometric literature has to date been somewhat sparse, although the companion to this book, Biometrics – Advanced Identity Verification (published by Springer, ISBN 1-85233-243-3) offers a comprehensive introduction to the subject. In retrospect, perhaps it isn't so surprising that this potentially exciting technology has got off to a slower than expected start.

A remedy was needed. A helping hand to assist those seeking to design and implement successful applications, whether as a technology supplier or end user. BANTAM has been conceived and introduced precisely for this reason. While described in general terms as a modeling language, BANTAM in fact goes much further, reaching into areas of procurement, training and fundamental program management. It is indeed a complete methodology for describing, designing, purchasing and implementing a biometric or related technology application. It may be used by systems integrators, suppliers, consultants, analysts and end users alike, as a common language for exchanging information and managing projects within this exciting area of technology. Furthermore, BANTAM has considerable value even after implementation, in areas such as training and systems maintenance, extending its reach into other organizational departments. The standard BANTAM documentation also provides an invaluable project archive which may be referenced and managed as part of an overall quality management program. The clarity and repeatability that BANTAM provides unleashes the potential for considerable cost savings throughout the entire project life cycle, especially with regard to resources. Indeed, BANTAM may be considered a significant step forward in the quest to realize the potential of biometric, token and related technologies in the context of personal identity verification.

After reading this book and experimenting with the BANTAM components and utilities provided on the accompanying CD-ROM, you will be ready and able to use BANTAM effectively across a broad range of projects dealing with biometric or token technology and related applications. You will find the methodology intuitive, easy to use and especially valuable when discussing aspects of process or systems design, both internally and with third parties. I sincerely hope that you enjoy reading this book and wish you every success in your future endeavors.

Julian Ashbourn

Contents

1. Introduction

It is interesting to note how the concept of application modeling has evolved in line with object oriented programming and rapid application development (RAD) techniques in general. In the early days of computing and application development, programmers would often spend considerable time liaising with users from within the business area in order to understand the requirement and translate it somehow into workable code. Sometimes of course, the programmer might interpret elements of the requirement slightly differently from the user representative, and sometimes the expediency of coding technique would take precedence over ultimate usability. To a certain degree it was inevitable, at least within the bounds of reasonable cost, that the user would have to accommodate the eccentricities of the computer and mould their processes accordingly. At least, this is how they saw the situation, and it was a situation which suited the programmers rather well. Computer programmers in those days were considered something of an elite and rare breed with whom you didn't argue, at least not if you ever wanted to see the realization of the planned application.

Time changes everything however, and nowadays every college student is a closet computer programmer, as sometimes are many end users who are the customers for today's application developers. There is thus a slightly different perception of what it takes to develop a typical application as might be used within a business environment. In this respect, RAD techniques have simplified much of the routine and time-consuming coding of yesterday with almost a drag-and-drop approach to development, utilizing objects and existing operating system APIs to streamline the task considerably. In addition, this approach promotes consistency and reusability of code. However, in counterbalance of this desirable state of affairs, we must acknowledge that many organizations and institutions have a hugely complex IT infrastructure, often as a result of different vintages of applications trying to talk to each other. The interfaces and dependencies within such an environment are sometimes complex and less easily understood than any individual application. Developing a new application to fit seamlessly within the overall picture can therefore be a nontrivial task on occasion, in spite of the gains made with contemporary development languages and environments. To complicate the picture further, we have the introduction of new peripheral technologies which introduce their own idiosyncrasies of operation. Just how much easier the programmer's job has become is therefore a debatable point. On the one hand, he or she can undoubtedly knock out enough code to produce working prototypes in a comparatively short time – a matter of a few hours sometimes. However, debugging, refining and porting the resulting application to the plethora of data sources and interfacing applications and services often required in order to deliver the business benefit sought, can sometimes take a considerably longer. In addition, this very complexity opens the door for potential misunderstandings among the user and developer community, to the degree that the finally delivered application may or may not provide the originally requested functionality. This problem may be further exacerbated by the business requirement having changed within the prolonged development cycle.

For reasons such as those described above, various project management methodologies have been introduced to encourage an iterative approach to application development, with the production of prototypes, or even finished chunks of code appertaining to a particular function, in order that these can be refined and approved along the way. This is fine, but could be time-consuming if each iteration was poorly understood by those concerned and subsequently had to be re-addressed and re-presented, perhaps a number of times, before finally being approved. Such a methodology certainly brings structure to the whole process, but it tends to rely somewhat on the original requirement definition being correctly articulated in a manner which facilitates the design and build of the application. This may be easily achieved in the case of a small, conventional application development, but rather less so for a complex application that crosses many operational boundaries, or one which is introducing new technology or new concepts. In such a situation, the concept of 'modeling' the application using a defined and repeatable technique can be extremely helpful.

Taking the above point, it follows that such a modeling technique needs to be encapsulated in a language that may be easily understood by everyone in the chain, from end user, through analyst and consultant, to application developer. It would be beneficial if such a language included a graphical notation to illustrate concepts and relationships clearly and unambiguously. In addition, such a technique should be capable of clearly documenting the requirement in a way that promotes subsequent understanding, perhaps by different individuals, long after the event. In short, a modeling language is all about communication: communication of the original idea; communication of all the attendant processes from the user perspective; communication of the process and application logic; communication of the logical and functional design; and so on. This fundamental requirement of modeling, *i.e.*, communication, should not be forgotten by those using such techniques. There are, of course, established modeling techniques and languages in circulation for general IT use, and these have been generally well received, indeed, leading to the creation of a modeling language industry in itself. There are conferences on modeling languages, a host of associated books and publications, various internet sites and of course, a number of vendors selling related goods and services around modeling. In addition, there are modeling software packages which seek to automate much of the complexity around modeling. These are based upon established modeling languages and use the concept and notation of the relevant language accordingly.

1.1 Why this book exists

As indicated previously, there are already established modeling languages available which have been well accepted and are practiced widely, so why introduce something different? Couldn't we simply recommend the use of an existing language? Well, certainly, we could indeed recommend the use of an existing language. However, there are two reasons why this course is not being followed in this instance.

Firstly, as already hinted at, certain existing languages seem to have had as their goal the generation of an all-encompassing modeling language, becoming increasingly complex in the process. This somewhat defeats the object of the original concept. If a given language is so complex that users have to be extensively trained before they can make sense of it, then it is hardly likely to provide much benefit to others who are unfamiliar with it. Furthermore, some would say that if you adhere to these languages to the letter and follow all of the required steps along the way, then you may just have well spent the time writing and documenting the application, which may actually have been a shorter task in some circumstances! In addition, if one uses the recommended software tools, there is a not inconsiderable cost associated with doing so. This may be considered worthwhile in certain circumstances and clearly it is a question of horses for courses as to the use of established techniques, but, in general, it may be fairly considered that adopting such techniques will be nontrivial for many organizations. One also has to ask whether the use of such sophisticated languages is actually simplifying or, indeed, complicating the overall scenario? If the latter is true, then this must be taken into consideration.

Secondly, the area of biometrics and token technology has its own peculiarities around typical end users, applications, operation and systems design. In many ways, it is an ideal area in which to use a modeling language, providing such a language can be kept simple and intuitive, facilitating its use throughout the project cycle, from original concept to delivered application, and among all those concerned along the way. In this respect, clarity and ease of use are primary concerns, followed by the applicability of the technique to the associated documentation and best practices within this sector. Using an established modeling language, while certainly possible, would perhaps prove a little clumsy and unintuitive in this instance. What is needed, is something altogether more elegant and streamlined which can be quickly adopted and understood by all those involved in this particular area of technology. This is where BANTAM comes in. BANTAM is an acronym for Biometric and Token Technology Application Modeling Language, and, as the name suggests, it is a small, light weight modeling language and methodology that may be easily and quickly adopted in order to bring consistency and clarity to systems design in this area.

One of the characteristics of biometric and token technology from an implementation point of view, is that although there has been much good work undertaken in the area of standards, many of the available devices are fairly proprietary in nature. Similarly, many of the available software utilities are equally proprietary. This sometimes leads to complications when designing a sophisticated system, especially when the end user is trying to put together a meaningful specification with which to go out to tender. Often, there is relatively little information available from vendors as to the precise technical operation of systems components, making it difficult to suggest a design that may be realistically provided or maintained in a vendor transparent manner.

One way of addressing this situation is to design the system conceptually and according to the desired logical process flow, and then invite potential suppliers to explain how they will meet the requirement. The problem here however is one of interpretation, as it is unlikely that all the parties concerned will understand the problem, or the potential solution, in quite the same way. The end user will veer towards the perceived logical process. The device vendor will veer towards an implementation that suits their standard product, and third party suppliers will often just state the interface requirements of their particular components. To complicate matters further, there is often a tendency for potential suppliers to agree to requirements that are ill-defined to start with, leading to the inevitable crisis further down the line. The net result of systems designed in this way can be a rather unsatisfactory sale for the vendor and a headache for the user when the system doesn't live up to expectations or fulfill the original requirement as the user saw it. This is assuming of course that the system works as intended in the first place and doesn't cause wider network related problems. Of course, it is not all doom and gloom and there are many well defined, and beautifully designed and configured systems working reliably out there, but this is not easily achieved when working with relatively new technology for which there is not necessarily a strong understanding among either users or systems integrators.

BANTAM is designed to help in such a situation, by providing an easily understood and highly portable methodology which can be utilized at every stage, from initial aspiration, through formal requests for proposals, to process definition and actual systems design. Furthermore, the consistent documentation and clarity of presentation that BANTAM provides, makes it easy for everybody involved to speak the same language and maintain the same understanding as the project progresses through its various stages. One of the key strengths of BANTAM lies in its iterative capability, whereby a relatively simple high level application logic map may be enhanced with the addition of specific attributes, or references to lower level detail maps, while maintaining its original clarity. All BANTAM documents are thus related and may easily be used in discussion in order to progress towards a final and agreed systems design on which to base actual development and implementation. Even beyond implementation, BANTAM documents may be used for support and maintenance purposes and in user training.

BANTAM is also complementary to existing standards and best practice as object attributes may reflect common protocols and functions accordingly. When viewed in this light, it is easy to understand the potential strength of BANTAM and how it can act as the glue to draw together all the best elements of contemporary systems design, while remaining intuitive and easy to use for all concerned. This is precisely why the BANTAM modeling language and associated methodology has been developed and is being presented to you in this book. It is hoped that as either a user or practitioner in the field of biometrics and token technology, you will welcome the BANTAM philosophy and incorporate it as a standard way of articulating and developing systems design requirements. Indeed, after a little while, you will almost certainly find that you can extend the concept into many related areas, making BANTAM an essential part of your systems tool kit. You should also find that the use of BANTAM has the potential to save considerable amounts of time in the context of your biometric project. If we view time as being money, there is a direct value equation here, which could be significant in many instances.

In conclusion, a proper modeling language for biometric and token technology is probably long overdue. BANTAM effectively addresses that need, but goes even further to provide a complete framework within which to specify, design, implement and manage related systems.

1.2 Who is this book for?

One of the strengths of BANTAM is its portability across different elements of the project and project life cycle. Similarly, it is equally portable across different user profiles. This usability is a very important aspect of the methodology, as one of the primary aims of any modeling language is communication across boundaries of expertise. BANTAM takes this concept even further, as it is designed to be used by anyone who has an involvement with the project, whether they be the end user, the systems consultant, application developer or general administrator. The following give some idea of the potential scope of use.

Business champion

By business champion, we mean the individual or department who is sponsoring the project. This will often be the same individual or group who developed the original idea, usually in response to a specific problem or situation. The business champion will need to be able to articulate the requirement in a consistent and unambiguous manner in order to advise analysts, consultants and potential vendors. Furthermore, they will have to absorb a certain amount of technical detail (albeit at a high level) in order to be able to present the concept to others, perhaps for reasons of funding, general information or other associated requirements. The clarity and order that BANTAM brings to the project will be of great assistance to the business champion throughout the project life cycle and even beyond, as he or she may often be called upon to describe the project, its operation and benefits to other departments.

Business analyst

The business analyst will be responsible for identifying the current operational processes and designing the new processes to be used as a result of incorporating biometric and token technology. They must not only understand this in detail, but be capable of articulating it clearly to others. BANTAM is the business analyst's friend in this respect, as it provides the vehicle for clarity and consistency. Even better, it does so in a way which is easily understandable by everyone with whom the business analyst must liaise in order to capture the necessary existing process detail and test the proposed new process thinking. In fact, the business analyst will no doubt discover additional benefits to BANTAM the more it is used, as it will promote a working methodology that may be ported to various projects within the enterprise.

Consultant

Consultants used on the project may be internal or external and may be in the form of one or several depending upon the size and scope of the project. Part of the consultant's task is to understand the big picture and to be able to advise on a given element accordingly. For this, the consultant needs a concise way of describing any particular function within the broader context. The various BANTAM documents will be of significant value here, allowing the consultant to drill down to the finest level of detail when necessary, but always within a clear, easily understandable framework. Close liaison with others on the project is also a hallmark of the consultant's role, and BANTAM will facilitate pertinent discussions with a common understanding among the various parties.

Application developer

The application development team will no doubt be fully conversant, with both the application development tools at their disposal and the underlying systems architecture of the targeted deployment site (if they are an external team, naturally this information will need to be provided). They will also have a pretty good idea of how to write efficient code and fine-tune the application for the best performance with regard to databases *etc.*. However, what they won't necessarily have is a thorough understanding of the desired operational processes or, indeed, of the finer points of operation of the anticipated biometric and related devices. This information will need to be gleaned from discussion with other parties as appropriate. A perennial issue in this respect is that application developers typically speak a different language from business analysts and consultants and need to arrive at a common understanding of what is desired and what is possible.

This of course cuts both ways. The business analysts and consultants need to clearly articulate the requirements and the operational logic behind them, and the application developers need to articulate clearly why one approach is better than another programmatically, and what the consequences and compromises might be. For this, both sides need a common language with which to demonstrate concepts and logical flow. Enter BANTAM, which can serve from 'diagrams on cigarette cartons' to finished design documents and everything in-between.

Project manager

The project manager will no doubt have a preferred methodology and software tool for managing the overall delivery project and attendant milestones. He or she will also need at least a top level understanding of the systems design issues and operational processes, in order to place things into context. This is important when understanding the consequence of slippage or balancing priorities at a given stage within the overall project. The use of BANTAM facilitates this understanding. In addition, the BANTAM documentation can usefully complement the various project timeline and milestone charts, providing a further dimension of understanding for the project archives.

Support engineer

The often long-suffering support engineer is supposed to be able to solve almost any systems related problem almost immediately, regardless of the cause or operational state of the equipment concerned. Sometimes they can work miracles. At other times they can do little more than patch up an ailing system. Sometimes there is nothing they can do, short of replacing major chunks of the system. In order for them to do their job at all, they must have the necessary information about the underlying systems architecture, the application in question and any relative interfaces and dependencies. If this information is out of date, incomplete or in any way weak, then their ability to operate effectively is correspondingly impaired. It would help the support function a great deal to have this sort of systems related information captured within the BANTAM documentation. Indeed, a set of these documents could usefully be provided to the systems support function as a matter of routine for all such applications. The systems support personnel may even like to adopt the BANTAM technique for their own documentation and training purposes, as its clarity, conciseness and repeatability are well suited for mapping out elements of systems infrastructure in an intuitive, easy-to-use manner. It would also facilitate discussion with users where appropriate.

Systems operator

The systems operator needs to understand more than just the functionality of the front end software in order to be truly effective. He or she should also have a rudimentary understanding of how the overall system is functioning, especially with regard to the important areas of enrollment and authentication. What better way of describing this than with the same documents used for conceiving and designing the system itself? Using the BANTAM documentation in order to train the systems operators ensures no ambiguity between the actual system and their interpretation of it. It will also help them to understand what is happening around a typical transaction and how this might affect the perceived performance.

Systems trainer

In addition to training the systems operators, there is also the tuition of everyday users to consider. Depending upon the scale of the operation, this may be a significant task requiring a dedicated resource. In such a case, this training resource must itself acquire a thorough understanding of the system and its functionality, including the architectural design and its affect on performance. They must subsequently be able to articulate elements of this to the user population, ensuring that the training activity sits comfortably with the system as implemented and that it is suitably effective. The BANTAM documentation will obviously be extremely useful for training the trainers. However, subsets of it may also be useful for the trainers to adopt within their own training programs for the system users. Utilizing the same language naturally promotes consistency and reduces the possibility of misunderstanding when passing information between parties.

Program manager

I have included program manager as distinct from project manager, as the two roles can be notably different under certain circumstances. While the project manager will be primarily concerned with delivering the implementation on time and to budget, the responsibilities of the program manager may extend far beyond this and for the effective life of the implemented solution. The program manager will often be responsible for orchestrating both the initial success of the chosen solution, and for an ongoing process of continual improvement accordingly. He or she must naturally have a sound understanding of how and why the project was conceived, how it was designed and implemented, and how it functions on a day-to-day basis. They will also need to understand the variability of components and how they might be fine-tuned to affect performance, as well as the extensibility of the system overall. The BANTAM methodology and associated documentation will be of great benefit in this context, as the program manager can use the technique extensively and throughout the life of the project, for overall management purposes. Another part of the program manager's responsibility often lies around the overall 'marketing' of the project. In this context the BANTAM technique will also prove extremely valuable with its easily understood documentation and concepts.

Some generic examples have been given above of how different individuals within a typical project structure might benefit from using BANTAM. The reader will have noticed that a consistent and key element of this benefit lies around communication. Clear, unambiguous communication of requirements, system design and other elements can save considerable time and resources within any medium-to-large scale project or, indeed, within any project dealing with specialist devices and concepts. Not only does this make working on the project a pleasure for all concerned, but it removes the likelihood of catastrophic misunderstanding or misinterpretation of terminology or intent. In addition, the use of a methodology such as BANTAM facilitates the provision of a clear audit trail, covering conception, design and implementation, for archive and quality assurance purposes.

The adoption of BANTAM as a working methodology within the enterprise for biometric, token technology and related applications is thus enthusiastically recommended, as the small commitment required in learning the technique will surely be repaid many times over, even within the very first project to which it is applied. Who is this book for? It is for you, dear reader, as an interested party or practitioner, seeking to design and implement systems with an appropriate degree of structure, logic and consistency to enable successful conclusions – every time.

2. Technology overview

In order to place BANTAM in context, it is necessary to provide a brief overview of the technologies for which it has been designed. This is complicated by the fact that technologies such as biometrics and token technology are particularly fast moving, both in their underlying scientific development and in their applications. This section will consequently represent a snapshot of current thinking in this respect: it cannot predict the future with any certainty. However, the concept of BANTAM is such that it may easily be ported across complementary technologies where applicable, and the enthusiastic user should have absolutely no difficulty in doing this. Furthermore, it is somewhat inevitable that BANTAM will be reviewed from time to time, and possibly extended, in order to incorporate technological developments as necessary. In the meantime, a brief overview of current biometric and token technologies follows.

2.1 Introduction to biometrics

Biometrics are best defined as measurable physiological and/or behavioral characteristics that can be utilized within an automated system in order to verify the identity of an individual. They include fingerprints, retinal and iris scanning, hand geometry, voice patterns, facial recognition and other techniques.

It is tempting to think of biometrics as representing a futuristic technology that we shall all be using together with solar powered cars, food pills and other fiendish devices some time in the future. This popular image suggests that they are a product of the late twentieth century computer age. In fact, the basic principles of biometric verification were understood and practiced somewhat earlier - thousands of years earlier to be precise, as our friends in the Nile valley routinely employed biometric verification in a number of everyday business situations. There are many references to individuals being formally identified via unique physiological parameters such as scars, measured physical criteria or a combination of features such as complexion, eye color, height and so on. This would often be the case in relation to transactions in the agricultural sector, where grain and provisions would be supplied to a central repository, and also with regard to legal proceedings of various descriptions. Of course, they didn't have automated electronic biometric readers and computer networks, and they certainly were not dealing with the numbers of individuals that we have to accommodate today, but the basic principles were similar.

Later, in the nineteenth century, there was a peak of interest as researchers into criminology attempted to relate physical features and characteristics with criminal tendencies.   This resulted in a variety of measuring devices being produced and much data being collected. The results were not conclusive but the idea of measuring individual physical characteristics seemed to stick and the parallel development of fingerprinting became the international methodology among police forces for identity verification. The absolute uniqueness or otherwise of fingerprints is sometimes debated, and the criteria that different countries employ to verify a fingerprint varies across the globe with a greater or lesser number of minutiae points required to be matched. Added to this is the question of personal interpretation which may be pertinent in borderline cases. Nevertheless, this was the best methodology on offer and still the primary one for police forces today, although the matching process is very often automated these days.

With this background, it is hardly surprising that for many years a fascination with the possibility of using electronics and the power of microprocessors to automate identity verification had occupied the minds of individuals and organizations in both the military and commercial sectors. Early biometric readers were perhaps a little crude and rather expensive, but continuous evolution has refined them to the point where they have become reliable, relatively low cost and easily deployed devices, with many potential applications in government, the military, academia and industry. In addition, the last decade has seen the biometric industry mature from a handful of specialist manufacturers struggling for sales, to a global industry shipping respectable numbers of devices and poised for significant growth as large-scale applications start to unfold. Currently popular biometric methodologies include the following:

Fingerprint verification

There are a variety of approaches to fingerprint verification. Some of them try to emulate the traditional police method of matching minutiae, others are straight pattern matching devices, and some adopt a unique approach all of their own, including moiré fringe patterns and ultrasonics. Some of them can detect when a live finger is presented, some cannot. There is a greater variety of fingerprint devices available than any other biometric at present.

Potentially capable of good accuracy (low instances of false acceptance), fingerprint devices can also suffer from usage errors among insufficiently disciplined users (higher instances of false rejection) such as might be the case with large user bases. One must also consider the transducer/user interface and how this would be affected by large-scale usage in a variety of environments. Fingerprint verification may be a good choice for in-house systems where adequate explanation and training can be provided to users and where the system is operated within a controlled environment. It is not surprising that the workstation access application area seems to be based almost exclusively around fingerprints, due to the relatively low cost, small size (easily integrated into keyboards) and ease of integration.

Hand geometry

As the name suggests, hand geometry is concerned with measuring the physical characteristics of the user's hand and fingers, from a three dimensional perspective in the case of the leading product. One of the most established methodologies, hand geometry offers a good balance of performance characteristics and usability. This methodology may be suitable where there are larger user bases or users who may access the system infrequently and may therefore be less disciplined in their approach to the system. Accuracy can be very high if desired, while flexible performance tuning and configuration can accommodate a wide range of applications. Hand geometry readers are deployed in a wide range of scenarios, including physical access control and time and attendance recording, where they have proved particularly popular.

Voice verification

This is a potentially interesting technique bearing in mind how much voice communication takes place with regard to everyday business transactions. Some designs have concentrated on wall-mounted readers while others have sought to integrate voice verification into conventional telephone handsets. While there have been a number of voice verification products introduced to the market, many of them have suffered in practice due to the variability of both transducers and local acoustics. In addition, the enrollment procedure has often been more complicated than with other biometrics leading to the perception of voice verification as unfriendly in some quarters. However, much development work has been undertaken in this context and voice verification may be a viable technique for many applications.

Retinal scanning

This is an established technology where the unique patterns of the retina are scanned by a low intensity light source via an optical coupler. Retinal scanning has proved to be quite accurate in use but it does require the user to look into a receptacle and focus on a given point. This is not particularly convenient if you wear spectacles or if you have concerns about intimate contact with the reading device. For these reasons retinal scanning has a few user acceptance issues, although the technology itself can work very well. The methodology has proved to be reliable in certain high security applications where the user acceptance issue is less of a concern.

Iris scanning

Iris scanning is undoubtedly the less intrusive of the eye related biometrics. It utilizes a fairly conventional CCD camera element and requires no intimate contact between user and reader. In addition it has the potential for higher than average template matching performance. It has been demonstrated to work with spectacles in place and is one of the few devices which can work in identification mode (as opposed to 1-1 verification mode). When deploying iris scanning devices, the environment will typically need to be taken into consideration.

Signature verification

Signature verification enjoys a synergy with existing processes that other biometrics do not. People are used to signatures as a means of transaction related identity verification and would mostly see nothing unusual in extending this to encompass biometrics. Signature verification devices have proved to be reasonably accurate in operation and obviously lend themselves to applications where the signature is an accepted identifier. Curiously, there have been relatively few significant applications to date in comparison with other biometric methodologies, with correspondingly fewer vendors.

Facial recognition

This is a technique which has attracted considerable interest, but whose capabilities have often been misunderstood. Exciting claims have sometimes been made for facial recognition devices which have been difficult, if not impossible, to substantiate in practice. It is one thing to match two static images (all that some systems can actually do), it is quite another to detect and verify unobtrusively the identity of an individual within a group (as some systems claim). To date, facial recognition systems have fallen short of expectations in practical applications. However, progress continues to be made in this area and if the technical obstacles can be overcome, we may eventually see facial recognition become a primary biometric methodology.

Vein geometry

This is a technique which seeks to identify the vein patterns in the back of the hand or wrist. This methodology has actually been around for several years, but has recently enjoyed a resurgence of interest, with some new products entering the market. It is potentially quite user friendly, although it may be rather early to draw conclusions around relative performance.

There are other biometric methodologies including the use of scent, ear lobes and various other parameters. While these may be technically interesting, they are not considered at this stage to be workable solutions in everyday applications. Those listed above therefore represent the majority interest at this time.

With regard to the operation of biometric devices, while individual devices and systems have their own operating methodologies, there are some generalizations one can make as to what typically happens within a biometric system's implementation.

Obviously, before we can verify an individual's identity via a biometric we must first capture a sample of the chosen biometric. This sample is referred to as a biometric template and represents the reference data against which subsequent samples provided at verification time will be compared. A number of samples are usually captured during enrollment (typically three) in order to arrive at a truly representative template via an averaging process. The template is then referenced against an identifier, in order to recall it ready for comparison with a live sample at the transaction point. The enrollment procedure and quality of the resultant template are critical factors in the overall success of a biometric application. A poor quality template will often cause considerable problems for the user, often necessitating a re-enrollment.

Template storage is an area of interest, particularly with large-scale applications which may accommodate many thousands of individuals. The possible options are as follows:

1. Store the template within the biometric reader device.

2. Store the template remotely in a central repository.

3. Store the template on a portable token such as a chip card.

4. Store the template in more than one location.

Option 1, storing the template within the biometric device, has both advantages and disadvantages depending on exactly how it is implemented. The advantage is potentially fast operation as a relatively small number of templates may be stored and manipulated efficiently within the device. In addition, you are not relying on an external process or data link in order to access the template. In some cases, where devices may be networked together directly, it is possible to share templates across the network.

The potential disadvantage is that the templates are somewhat vulnerable and dependent upon the device being both present and functioning correctly. If anything happens to the device, you may need to re-install the template database or possibly re-enroll the user base.

Option 2, storing the templates in a central repository or 'directory', is the option which will naturally occur to IT systems engineers. This may work well in a secure networked environment where there is sufficient operational speed for template retrieval and comparison to be invisible to the user. However, we must bear in mind that with a large number of readers working simultaneously there could be significant data traffic, especially if users are impatient and submit multiple verification attempts. The size of the biometric template itself will have some impact on this, with popular methodologies varying between 9 bytes and 1.5k. Another aspect to consider is that if the network fails, the system effectively stops unless there is some sort of additional local storage. This may be possible to implement with some devices, using the internal storage for recent users and instructing the system to search the central repository if the template cannot be found locally.

Option 3, storing the template on a token, is an attractive option for two reasons. Firstly, it requires no local or central storage of templates (unless you wish it to) and, secondly, the user carries their template with them and can use it at any authorized reader position. However, there are still considerations. If the user is attracted to the scheme because he believes he has effective control and ownership of his own template (a strong selling point in some cases) then you cannot additionally store his template elsewhere in the system. As a result, if he subsequently loses or damages his token, then he will need to re-enroll. Another consideration may be unit cost and system complexity if you need to combine chip card readers and biometric readers at each enrollment and verification position.

Option 4, storing the template in more than one location. If the user base has no objection, you may wish to consider storing the template both on a token and elsewhere within the system. This could provide fast local operation with a fallback position if the chip card reading process fails for any reason or if a genuine user loses their token and can provide suitable identity information.

Your choice of template storage may of course be dictated to some extent by your choice of biometric device, as certain devices offer greater flexibility than others in this respect.

Having considered the template capture process, let's now turn our attention towards subsequent verification. The verification process requires the user to claim an identity by either entering a PIN or presenting a token, and then to verify this claim by providing a live biometric to be compared against the claimed reference template. There will be a resulting match or no match accordingly. A record of this transaction will then be generated and stored, either locally within the device or remotely via a network and host (or indeed both). With certain devices, you may allow the user a number of attempts at verification before finally rejecting them if the templates do not match. Setting this parameter requires some thought. On the one hand, you want to provide every opportunity for a valid user (who may be having difficulty using the system) to be recognized. On the other hand, you do not want impostors to have too much opportunity to experiment. With some systems, the reference template is automatically updated upon each valid transaction. This allows the system to accommodate minor changes to the user's live sample as a result of ageing, local abrasions *etc.* and may be a useful feature when dealing with large userbases.

We have used the term 'verification' and, indeed, the majority of available devices operate in verification mode. This means that an identity is claimed by calling a particular template from storage (by the input of a PIN or presentation of a token) and then presenting a live sample for comparison. Thus a simple one-to-one match is undertaken, which may be performed quickly and generate a binary yes/no result. Certain devices can also operate in identification mode, whereby the user simply submits his live sample and the system attempts to identify him within a database of templates. Obviously this entails a more complex one-to-many match, which may generate a multiple result according to the number and similarity of stored templates and the matching threshold criteria used. While certain devices have been found to work well in this manner with small databases of tens of users, the situation can become quite complicated with databases of even a few hundred. The mathematical probability of finding an exact unequivocal match within such a database is rather slim, and a database of many tens of thousands of users presents an even larger challenge. Currently, iris scanning is the only biometric methodology that can work acceptably in identification mode with sizable databases.

In terms of performance metrics, the terms 'false accept', 'false reject', and 'equal error rates' are dominant in device specifications and biometric literature. False accept rates (FAR) indicate the likelihood that an impostor may be falsely accepted by the system. False reject rates (FRR) indicate the likelihood that the genuine user may be rejected by the system. This measure of template matching can often be manipulated by the setting of a threshold which will bias the device towards one situation or the other. Hence one may bias the device towards a larger number of false accepts but a smaller number of false rejects (user friendly) or a larger number of false rejects but a smaller number of false accepts (less user friendly), the two parameters being mutually exclusive. Somewhere between the extremes is the equal error point where the two curves cross and which may represent a more realistic measure of performance than either FARs or FRRs quoted in isolation. This will be expressed as the equal error rate. These measures are usually expressed in percentage (of error transactions) terms.

However, the story is not quite as simple as this, as the quoted figures for a given device may not be realized in practice for a number of reasons. These will include user discipline, familiarity with the device, quality of templates, user stress, individual device condition, the user interface, network performance and other variables. The popularly quoted biometric device performance metrics should therefore be viewed as a rough guide to performance under typical conditions, rather than a necessarily obtainable performance. This situation is not because vendors are trying to mislead, but because it is almost impossible to give an accurate indication of how a device will perform in a limitless variety of real world conditions. In addition, there are other parameters which should be taken into account, including typical transaction time, database performance and so on.

From the above, it will be clear that the use of a biometric device within a broader system entails an understanding of many parameters and possible options which can affect overall system performance. Describing and documenting these variables in relation to the overall requirement has historically been something of a challenge. The use of BANTAM will bring a welcome degree of clarity and consistency to this particular problem.

The above represents a necessarily brief overview of biometric technology for the purposes of this book. For more detailed information, please refer to the established literature on the subject (Biometrics – Advanced Identity Verification published by Springer, ISBN 1-85233-243-3 is recommended).

2.2 Introduction to token technology

Tokens for electronic systems have been around in various forms for a number of years and many millions are in everyday circulation, in the form of bank and credit cards, access control cards and other implementations. The technology used varies between relatively low tech magnetic stripe, to radio frequency contactless cards and chip cards with their own integral processing units.

As indicated, tokens have been used in a wide variety of applications from tagging animals in agricultural scenarios, through bank and identity cards, to unique ID tokens on automobiles, and many others. However, for the purposes of this overview, we shall concentrate on the types of tokens typically used in association with personal identity related applications, and which therefore have a synergy with biometrics. Of course, BANTAM may be used as a methodology for any application where it would be useful to describe the relationships and interactions between tokens and the wider application.

Tokens used for personal identity and related purposes have assumed a wide variety of shapes, sizes and utilized technology. Mostly, they have taken the physical form of a plastic card, however some of them have not been cards at all but specially shaped tokens to fit on key fobs, and other variations on a theme. Returning to cards, the most basic coded card technology was probably the punched hole card which simply had a series of holes punched within an area to form the code. Obvious limitations of this idea were that the range of possible codes was a product of hole size and available real estate for the coding area which dictated a limited range of codes. Furthermore, they were not particularly secure as duplicate cards could be made fairly easily by copying the pattern of punched holes. However, they were cheap to manufacture and the associated card readers were relatively simple and reliable using either mechanical sensors or transmitted light to read the code.

A methodology not too dissimilar was that of infrared cards. IR cards employed a mask between layers of the card which was coded either to allow or block the passage of infrared light. This technology had certain advantages in that the code was now invisible to the user, making duplication extremely difficult unless you had access to sophisticated machinery, and the cards themselves were quite robust. There was a slight disadvantage in that the readers were relatively expensive compared with contemporary card access technologies, although the infrared approach did prove to be reliable with large numbers of readers and cards deployed across a broad variety of applications.

A somewhat less successful card technology was barium ferrite. In simple terms, a barium ferrite card may be thought of as a layer of magnetic material sandwiched between the plastic layers of the card. This material could then be polarized by a high intensity electromagnetic field in order to produce a set code pattern. The barium ferrite readers contained an array of coils in set locations which would detect the coded pattern. This required quite careful alignment of the card and reader, which could sometimes prove problematic in practice. In addition, there were all sorts of stories circulating about barium ferrite access control cards affecting other magnetic media, such as credit cards, when placed in close proximity, for example in a purse or wallet. No doubt some of these stories were exaggerated, but mud tends to stick and barium ferrite cards became increasingly less popular as other more reliable methods became available.

Many people would associate the magnetic stripe card with bank and credit cards, access control and related applications, and certainly very large numbers of these cards have been used in this context. They have the advantage of being inexpensive, easy to code and easily read by inexpensive readers. A property of magnetic stripe cards often referred to is coercivity Coercivity may be thought of as a measure of how easily the code on the magnetic stripe may be accidentally (or otherwise) erased. Initially, the standard was for low coercivity and it was soon discovered that the coding on low coercivity cards was not particularly robust and could be easily altered by the presence of magnetic fields. High coercivity cards offer an improvement in this respect and the majority of magnetic stripe cards you come into contact with today will be of this persuasion. In a magnetic stripe card reader, a reading head is employed which is not unlike the head in a domestic tape recorder and similarly reads the polarized encoding of particles on the magnetic tape. This involves physical contact and high usage will eventually result in wear to the reading head. Ruggedized versions of magnetic stripe card readers have been produced, including insertion type readers, as used on ATM machines for example, where the reading head and mechanism is more protected.

One of the most reliable and robust card technologies is undoubtedly Wiegand. Wiegand cards incorporate two rows of tiny wire particles representing binary zeros and ones. However, the wire itself is very special with a soft magnetic core contrasting against a hard magnetic shell. The physical placement and polarization of these wire particles provides for a broad range of codes in a read-only format. Wiegand cards cannot be reprogrammed and are therefore very secure. Both cards and readers are very robust and can be utilized in harsh environments if necessary. The Wiegand reader is essentially non-contact in that the cards are swiped through a focused magnetic field and is typically hermetically sealed against dust and moisture, providing reliable reading in a variety of situations and environments. Wiegand card readers come in both swipe and insertion format, although the swipe readers are by far the most popular.

Another technology that springs to mind when discussing cards is bar coding. There are various forms of bar codes, the most popular perhaps being the linear code, as often seen attached to products for labeling purposes, and the two-dimensional bar code, which is rather more sophisticated and can hold considerably more data within a given physical area. Bar codes do lend themselves to individual identity applications in that there are a large range of possible codes, even in linear bar codes, however, as linear codes are represented by a printed pattern they are easily copied and cannot be considered secure. This may be a little more difficult with two-dimensional codes, but not impossible. One way of getting around this problem is to obscure the code so that it is not visible to the human eye, but easily read by the infrared bar code scanner. This may perhaps be regarded as a step forward, except that bar code reading and writing equipment is readily available and therefore duplicates may perhaps be easily manufactured. However, one would probably not choose this specifically as a high security technology. A common approach is to add bar code reading capability as an extra function to an existing token, enabling multifunctional tokens within the enterprise. Two-dimensional bar codes are sometimes used purely to hold large amounts of encoded data on an identity card. For example, it would be perfectly possible to encode an identity badge photograph into an associated two-dimensional bar code, enabling tampered cards (where the photo ID has been changed) to be detected. Alternatively, we may choose to encode additional information into the bar code in order that it can be machine read where appropriate but not visible to the card holder under normal circumstances.

Proximity cards and tokens have been popular where a noncontact reading methodology is required. There are two potential advantages with proximity technology. One advantage is that as there is no physical contact between cards and reader, the reader itself can be made very durable for external use or even built in to another structure to make it relatively vandal-resistant. For example, a reader may be hidden beneath a ceramic surface such as an office wall. The second primary advantage, particularly with long-range proximity cards and readers is that the user need not undertake a physical process to read the card. As long as he or she is in range of the reader, the card will be read, providing a 'hands-free' capability. The proximity token may even be attached to a vehicle in order to activate a car park barrier or otherwise track vehicles entering a particular area. There are also some disadvantages with this technology, not least being cost as proximity cards and readers are typically more expensive than the contact card methodologies. In addition, physical placement of the reader can be important as in most cases the reader is emitting an rf (radio frequency) field into which the card is brought at the time of reading, both card and reader acting as an antenna. This field may sometimes be distorted by the close proximity of metallic objects, producing unwanted and sometimes variable effects. Proximity cards may be either active (with an integral battery providing power) or passive. Active cards offer enhanced performance and read range but have a finite life, after which they must be replaced, adding further to overall costs.

There have been other specialist card technologies, like optical cards for example, in addition to the popular types mentioned above, but let's move on to the more interesting area of smart cards or, as I prefer to call them, chip cards, which have been in development more or less since the early 1970s. As the name would suggest, chip cards incorporate a chip embedded into the plastic card. This chip may either be a memory chip or a more sophisticated processor chip, incorporating its own operating system and instruction set. Memory-only chip cards have become less expensive as larger numbers are deployed and they therefore represent a viable and interesting alternative to the other technologies mentioned as they can hold a much larger amount of data in a relatively secure manner. However, the chip card readers are somewhat more sophisticated and expensive and may not always be suitable for harsh environments. Processor chip cards are particularly interesting and can offer a range of functionality including advanced data encryption techniques and the promise of a truly multifunctional card where appropriate. To many, this multifunctional approach is the way forward for chip cards and although progress has perhaps been a little slower than anticipated to date, there is no doubt that this idea is of considerable interest and offers some interesting possibilities for both organizational collaboration and real value to the user. One of these possibilities lies around incorporating a biometric template into the chip card for personal identification purposes, or maybe using a biometric to release a private key stored within a secure area of the chip card. Indeed, the potential integration of chip cards and biometrics into a PKI (public key infrastructure) is currently of particular interest to many organizations, and there is much background work being undertaken in this area.

As one would imagine, there are many possible variations on a theme with regard to the use of tokens within personal identity authentication related applications, making a clear design statement all the more important for those seeking to incorporate such technology into broader projects. In this context, the use of BANTAM will pay dividends in matching the solution to the requirement and providing clear, unambiguous documentation throughout the project life cycle. Systems integrators will particularly benefit from the BANTAM technique, being able to clearly assign attributes to the various component objects within a token based system.

As with our biometrics overview, the above is necessarily very brief, and the interested reader will no doubt like to undertake a little further research into the subject, especially with regard to chip cards and multifunctional tokens, where the technology continues to develop and provide further opportunities for integration into a broad range of applications.

2.3 Relevance of IT

Having discussed biometric and token technology in general terms, it is important to consider how these technologies are typically deployed and integrated into the broader systems scenario. This in turn has a relevance to the sort of project and program management techniques used within the typical IT based environment and to how BANTAM fits within this broader picture.

When considered in isolation, a biometric or token reading device will appear to the onlooker as a self-contained entity, and it is easy to become pre-occupied with individual device performance metrics and protocols. However, such devices offer little benefit until they are integrated into a larger system, which itself has specific objectives aligned with the originally perceived requirement. When the devices are integrated, then realized performance from the user perspective is a product of the total system performance, including that of any interfaces, databases and networks involved in the overall end-to-end design. Thus, an integral part of the design of any biometric or token based system (excepting true stand-alone devices), is the IT backbone which facilitates its use. An optimally designed system would therefore be designed as a whole, incorporating both the front end technology and the underlying infrastructure. This doesn't always happen in practice, especially with device vendor designed systems which tend to be focused on the front end technology, regarding the IT infrastructure as something they 'plug into' and which is the responsibility of the user. Often, the only systems related recommendations in this respect refer to the type of operating system and rudimentary memory requirements, with little reference to the broader systems architecture.

There is obviously a huge potential for misunderstanding within the above-mentioned scenario, which can, and often does, lead to suboptimal systems being installed, which never quite reach their potential, or deliver the expected performance. In order to guard against this situation, it is imperative that the host systems architecture is fully understood, including any relevant capacity and performance characteristics, before we consider the integration of any third party biometric or token components. Similarly, we need to understand how such components will perform when integrated into a given infrastructure, and whether we need to consider enhancements to the infrastructure in order to support the new application fully. Before we can discuss and reach conclusions on these issues, we need to understand them fully, and be able to articulate clearly the associated options and consequences in a way that all parties will understand. The use of BANTAM will facilitate just this approach and dramatically reduce the potential for misunderstanding accordingly.

To illustrate this scenario, imagine an application where we are aiming to use a network of 35 biometric readers within a government building complex. The network will be administered from a central point, and the readers will communicate via the existing LAN, with biometric templates stored in a central directory which is also used for network access control purposes and includes details for 2000 personnel. The manufacturer of the biometric reader has quoted performance figures measured in isolation, under laboratory conditions, and with the reader connected directly to a single high-performance personal computer. The biometric readers may be configured to use centrally stored templates, but the manufacturer has made no distinction within their claimed performance figures for this mode of operation. So how will the system perform in practice? In such a scenario, there are many questions that need to be asked. Firstly, is 35 the right number of readers, and how has this figure been arrived at? How many individuals are required to use each reader and at what times of day? Have assumptions been made about individual transaction times, and if so what are they? Are there network 'hot spots' at certain times of the day when data traffic is higher than average, and how does this fit with the anticipated peak use of the biometric readers? How is network performance and capacity currently measured, and what sort of contingency for growth is incorporated in the current configuration? Are there pending plans to modify the network in any way? What sort of database structure does the central directory sit on, and how scaleable is this? What sort of database redundancy and resilience is in place? Do the protocols in use support the exchange of biometric data? Will data encryption be used and, if so, what transactional overhead will this entail? We could pose many more questions, but from the above alone, we can see the importance of understanding the complete picture and being able to clarify the relevant points before finalizing a proposed system configuration. In order to provide this clarity, we need a working methodology which allows us to capture the relevant detail in an intuitive manner, at the same time generating a log of agreed systems related issues which may be subsequently addressed and logged as appropriate. In this way, we can incrementally build both an in depth systems understanding and the new application specification, until we are comfortable that what we have is a realistic and workable systems design, which is understood by all concerned and for which our performance expectations are equally realistic.

BANTAM provides us with such a methodology, in a particularly intuitive and easy to use manner, with the minimum of learning curves required in order to become familiar with the technique. Furthermore, the use of BANTAM may be acknowledged within conventional project management methodologies and systems, using the BANTAM documentation to confirm milestones, or indeed as milestones, in the case of process and systems analysis.

We have used a relatively straightforward example to illustrate the importance of looking at the system as a whole, and integrating the IT infrastructure into our overall systems design. This becomes even more important when considering a more complex system, such as might be the case with a mobile communications application, or a system based around internet transactions. Take the latter for example, the systems architecture around an internet based transactional system is likely to be quite complex, probably with a web server, applications server, back end databases, links to other systems, authentication and central directory services and, of course, the various firewall and routing components. Understanding exactly how all this fits together, including the various protocols and interfaces is nontrivial, before you even start thinking about your new application. This architectural picture needs to be clearly defined and documented in order to understand how new applications may be overlaid onto it. There will necessarily be a great deal of detail involved here, almost certainly too much for a third party consultant or systems integrator to understand in one pass. This is where a modeling language such as BANTAM pays dividends, as the complex larger picture can be broken down into more manageable chunks and clearly described and annotated. This, in turn, makes it easier to see how new components can be integrated and where new services are best located within the overall infrastructure. With clear, unambiguous architectural details to hand, the systems designer can start to design the new system with a high degree of confidence as to its operability and performance. Any weak points within the infrastructure will also be highlighted, enabling decisions to be made around enhancements where appropriate.

Within a large and complex system, the authorization and authentication services will also require careful consideration. With a biometric system for example, we shall have to make decisions as to where the biometric templates are stored and where the matching process takes place. If the templates are stored in a central directory and the matching engine also located centrally, then we shall want to understand the relevant communication protocols and exactly how data is passed between the various components. If biometric data is passed between the remote user interface and the central directory server, then we shall probably wish to employ some sort of strong encryption in order to protect the data in its journey across an untrusted network. But how exactly does this encryption work, precisely how and where does the decryption take place, and what transactional overhead does it introduce? Similarly, with respect to other token devices, we need to determine exactly how data is passed between the various system components, as well as the business logic involved in processing this information and the interfaces to other data sources. Naturally, we need to understand this level of detail, and with the help of BANTAM we can capture, articulate and map this information in an intuitive, easy-to-understand manner, no matter how complex the overall system is. Indeed, without such a tool it would be quite difficult to achieve suitable levels of clarity and consistency in this respect.

In conclusion, when considering an application which will use biometric or token technology, we simply cannot ignore the importance of the underlying systems infrastructure and the effect that this can have on both performance and reliability. It is therefore most important that we consider this aspect fully when designing our biometric or related application, and capture the relevant information in a manner that may be easily interpreted by all those involved with the project, whether they be suppliers, third party consultants, or in-house program management and development teams. Similarly, we need to understand and document all the relative interfaces, protocols and dependencies in order to ensure that our application is designed and configured in the optimum manner. An appropriate modeling tool will help facilitate all of this and will bring with it consistency and repeatability. BANTAM is just such a tool. Indeed, this philosophy is not limited to biometric and token technology but may be applied to any program dealing with the implementation of technology lead solutions. You may well find that BANTAM can be used effectively in many such programs, irrespective of the core technology being applied. Certainly the program management aspects of BANTAM, including the documentation, may be applied to virtually any project.

3. Introducing BANTAM

In the last chapter, we discussed the relevant biometric and token technologies and also considered the importance of the underlying systems infrastructure to our overall application. Several times, we mentioned BANTAM and the desirability of using a modeling language in order to articulate our requirements and design the application in an efficient manner. In this chapter, we shall take a closer look at BANTAM and how it fulfills these requirements.

Those familiar with traditional IT development and associated program management techniques will, of course, also be familiar with the concept of modeling languages, and might ask why we don't simply use an existing language for our biometric or token based application? Certainly, we could take that approach. However, existing languages don't necessarily lend themselves to the sometimes peculiar requirements and user scenarios under consideration here. In addition, they are perhaps a little too complex to be quickly adopted by all those in the chain, often requiring a steep learning curve of both the concept and, in some cases, an associated software application. BANTAM on the other hand, is (as its name suggests) designed to be lightweight, efficient and easily learned by all those involved in a typical project. Furthermore, and very importantly, it is designed not just for the benefit of application developers or systems analysts, but has real benefits in helping the end user to describe the requirement accurately and issue an unambiguous RFP (request for proposal) for potential suppliers. When suppliers respond to an RFP using the same technique, the end user can immediately understand what is being proposed and how it fits within the existing systems infrastructure. In this way, BANTAM becomes an extremely practical methodology, offering real and tangible benefits, while being particularly easy to understand and adopt across the enterprise.

3.1 The concept of BANTAM

BANTAM is actually quite straightforward in concept and revolves around a standard modeling notation and a portfolio of standard documentation. This makes it very easy for almost anyone to learn and use with the minimum of training. One of the primary objectives of BANTAM is to be usable at virtually all stages of the project, from original idea, through to managing the implementation and beyond into the realms of support and maintenance. It achieves this by using common notation and diagrams to describe processes and elements of systems design which can be used throughout.

The original concept may be mapped out with BANTAM and articulated accordingly. The business processes, both existing and future, may similarly be mapped. Specific technology elements and interfaces may be mapped out to show exactly how they should be configured. The underlying systems infrastructure may be mapped in order to show the relationship with the planned application. Specific functionality such as biometric authentication or a token transaction may be mapped both conceptually and architecturally. In fact, any concept, process, technical interface or data flow may be mapped using the simple BANTAM notation. When used in this way throughout the project, concepts, processes and system design elements may be easily articulated and understood between different parties using a common language.

The way of working described above ensures that opportunities for misinterpretation are minimized, thus providing greater efficiency. It also provides for consistent and intuitive project documentation which can be additionally used for archive purposes and subsequent support functions. This is an important point, as many systems end up being supported by persons other than the original project team, and with documentation which is ambiguous or incomprehensible, or worse still, no documentation at all, these people will struggle and support costs will escalate. Adopting BANTAM as a standard methodology will help to ensure that this does not occur.

In addition to the benefits described above, BANTAM reaches further into the realms of procurement, with the integration of BANTAM into the common procedures of requests for information (RFI) and requests for proposals (RFP). This is unusual among modeling languages and will be of particular value to government departments, military users and other large organizations who seek to standardize their procurement procedures wherever possible. Basing RFIs and RFPs on BANTAM maps will ensure that response documents are easily and consistently evaluated and that any deviation from the proposed specification is immediately noticed. This is a particularly pertinent point when dealing with areas of technology such as biometrics, chip cards and the like, where there still exists much scope for interpretation of both requirements and technology specification. The use of BANTAM helps to bestow clarity and consistency throughout all phases of the program, including procurement.

BANTAM is simple and straightforward in concept, and designed to be particularly easy to learn and use. With the use of this guide, there is no reason why everybody involved with a specific project shouldn't be able to use the BANTAM methodology within a few hours of introduction to it. However, there is more to BANTAM than the symbol notation. It is indeed a philosophy and way of working which, if adopted enthusiastically, will provide ongoing and substantial benefits across many areas of operation. This user guide will provide an introduction to the methodology, but it is with continued practice that the real strengths and benefits of BANTAM will emerge. Readers are thus encouraged to practice and experiment with the BANTAM tools and templates in order to develop their own understanding of how the methodology best fits within their particular organization or sphere of activity.

The standard BANTAM document set provides scope to describe virtually any element of a biometric or related project. There is perhaps a logical sequence to the preparation of the BANTAM maps, although there are no hard and fast rules on this point, just use them as required, when required. The Application Logic Map is a good place to start as it will help to focus thinking around the primary purpose of the application under consideration and how it might best be configured. Indeed, there may be iterations of this map as thinking evolves in this context. The Logical Scenario Map and Functional Scenario Map will help us to work through the operational processes and how these might be configured from a systems perspective. Again, there will probably be several of each of these maps in order to describe different elements of the application and how it should function. The Systems Architecture Map will be particularly important in order to test the feasibility of the proposed solution within the organizational systems infrastructure and highlight any areas which will need updating or improving accordingly. When the overall design and functionality have been ascertained, the Object Association Map will help the software developers and/or systems integrators to understand the relationships between different entities and ensure that all interfaces and dependencies are addressed. The concept of using consistent and repeatable maps throughout these different phases will quickly become understood and valued as the portability of definitions across departments becomes necessary. There may of course be other elements, unique to a particular project perhaps, which don't obviously fit into one of the predefined BANTAM categories. In such an instance, the Miscellaneous Definition Map may be used to illustrate and define the situation and functionality accordingly.

The next section will introduce the fundamental notation symbols – the graphical language of BANTAM. The notation is best viewed as a collection of building blocks which become meaningful when assembled together into the broader framework of one of the standard BANTAM documents. While at first glance they may appear rather simplistic and obvious, they nonetheless represent a potentially powerful way of describing any part of a typical biometric or related application. Indeed, BANTAM may be used for a broad variety of more generic systems applications if desired. However, the notation symbols should be used in context and qualified by further definition within the Map Explanatory Notes where applicable. The reader is therefore encouraged to read the user guide in its entirety before using the methodology in earnest. An in-depth understanding of the BANTAM document set and its application, will help to place the notation language in context.

3.2 BANTAM notation

This section will provide an introduction to the BANTAM notational symbols and their recommended use within the BANTAM documentation.

The user symbol

The user symbol is used to depict the user wherever he or she interacts with the application under consideration. This interaction may be directly with a systems component, such as a user interface or biometric capture device, or it may be more process related such as might be the case when mapping current or anticipated operational processes. For example, what actions does a user generally undertake within an existing process? Do they have a one-to-one dialog with another user in order to verify a particular situation? Are they required to move physically to another location in order to perform some function? Do they access a computer terminal or some specific piece of machinery? Do they make a telephone call? Or look at a CCTV monitor? Or check some paper credentials? How does this differ between different types of user? All of these may be captured within the standard BANTAM maps using the user symbol as appropriate.

Wherever a user action or reaction has some relevance to either the system or process being described, then we may utilize the user symbol within BANTAM to illustrate the instance. As with all the BANTAM symbols, we may add attributes to the user symbol in order to qualify the instance as a particular type, such as regular user, systems administrator, first time user and so on. If we wish to distinguish a user in this manner, then we may simply add text next to the symbol within the BANTAM map and further qualify the user type within the Map Explanatory Notes if required. Of course, the user symbol may also be used to depict an administration function such as trainer or enrollment officer.

The biometric device symbol

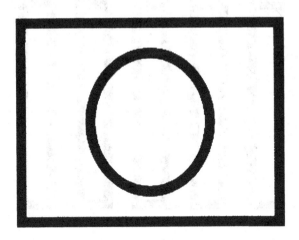

The biometric device symbol represents the biometric capture device, such as a fingerprint reader, hand geometry reader or an imaging device used for iris scanning or facial recognition for example. It is used within the BANTAM maps to depict a biometric device in relation to other components and functions. In a Logical Scenario Map for example, the biometric device may be simply shown in relation to the logical flow of events, whereas in a Functional Scenario Map, the biometric device might be shown together with its detailed interfaces and connections to other components as well as in relation to transactional functionality.

The biometric device symbol may also be used in conjunction with the object class symbol within an Object Association Map. In such an example, attributes and methods may be assigned to the biometric device via the object class symbol. For example, attributes might include free standing, integrated or perhaps a particular model designation. Methods might include capture data, process image, send data, encrypt data, connection handshake and so on.

Where a biometric device is particularly complex or perhaps of custom configuration, then it may warrant a map of its own in order to depict such customization and its impact on surrounding components. This device specific map may then be referenced via the BANTAM reference symbol from other higher level maps. Similarly, if there are specific performance or durability requirements, as may be the case for use within a hostile environment for example, then a separate map may be produced in order to capture these requirements. The biometric device symbol is therefore used to depict an instance of a biometric device within a particular map, whether it be process, function or architecture specific.

The template symbol

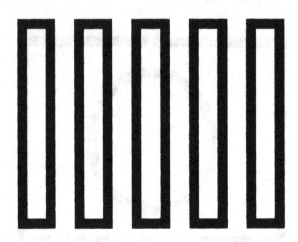

The template symbol is used to depict a biometric template within the appropriate BANTAM maps. Its use may therefore be quite widespread as it may be used at a high level within an Application Logic Map to show the presence of a template, or a more detailed level within a Functional Scenario Map to show exactly how a template is created, stored, retrieved or otherwise used. It may also be used within a Systems Architecture Map to show exactly where the template is stored and how it is accessed, or a Logical Scenario Map to show the relationship with the user and its place within the authentication transaction. It may even be used within an Object Association Map in order to depict its attributes and relationship with other system components.

The template symbol will often be used in association with other symbols such as the token symbol for example, or perhaps the database symbol. It will certainly be used in conjunction with the authenticate symbol in order to depict the mechanics of the authentication process.

From an application developer or systems integrator perspective, it will be important to understand details such as the template size, the file format under which it is manipulated, the compatibility with generic system components and so on. A specific BANTAM map may be produced in order to either request or provide such information and may be subsequently referenced within other maps if required. The associated Map Explanatory Notes may be used to highlight other attributes or provide additional detail around the use of the template under various scenarios. For example, it may be possible to use partial or subset templates for certain applications, or have a richer template, perhaps at the expense of storage size or matching algorithm performance. If such variations exist, or are possible for specific usage cases, then the BANTAM maps may be used to describe such variations and the impact they may have from an overall systems perspective.

The token reader symbol

The token reader symbol represents a token reader within any relevant BANTAM map. This symbol will no doubt be used in a variety of contexts, from high level definitions within an Application Logic Map to detailed descriptions within a Systems Architecture Map and many other areas besides.

Within an actual application, the token reader deployed will be of a specific type and configuration. In this respect, it may be worth producing a separate BANTAM map in order to show the detailed configuration, interfaces and operational logic of the token reader. This map may then be referenced by higher level maps as appropriate within the BANTAM documentation set. The token reader symbol may also be used in association with the object class symbol within an object association map if required in order to show the relationship with other system components.

As with other BANTAM symbols, the token reader symbol may be further qualified by appending attributes and methods if this helps to define a particular process within the overall application. Any such attributes or methods should be explained within the Map Explanatory Notes attached to each primary BANTAM map. In Logical and Functional Scenario Maps, the token reader symbol will often be depicted in association with the token symbol, either to show the logical flow of events around a specific transaction, *i.e.,* when exactly the token is used, or to define the processes around the use of the data on the token and the relationships between this data and other entities or processes within the overall system. One should also use the token reader symbol where appropriate, within the description of an integrated device such as a kiosk, user terminal, ATM machine or similar and where such interfaces need to be clearly defined.

The token symbol

The token symbol is naturally used to depict a token within the various BANTAM map documents. However, this depiction may take several forms depending on the context. Within Application Logic Maps or Logical Scenario Maps, the token symbol will typically be used to indicate the presence of the token within an overall process or transaction. Within Functional Scenario Maps, the token symbol may be used with appended attributes to describe the data associated with the token and how it is utilized or referenced within a given transaction. For example, if a biometric template is stored on the token and is subsequently transferred to a host system for verification purposes, then we can describe this clearly within the appropriate BANTAM maps. Similarly, data may be returned to the token in relation to a transaction and this will also need to be defined and described accordingly.

If the token is used for multiple purposes, some of which may not be directly related to system transactions, then we may consider producing a separate Miscellaneous Definition Map in order to describe this situation in detail. This may be the case, for example, where the token is used for manual visual identification, perhaps as a company employee card or even a national identity card. Alternatively, there may be other token holder information such as name, account number, expiry date and so on, when the token is perhaps used as a travel card, credit card, medical card or something similar. This can all be specified and described within the BANTAM documentation, right down to the manufacturing specifications of the token itself if required.

The token symbol will thus probably appear quite frequently within the BANTAM document set where the application in question uses tokens, whether they be a core part of the application or, indeed, in cases where existing tokens are to be integrated into the application.   .

The host application symbol

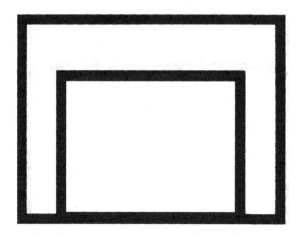

The host application symbol is used to depict the host application which is responsible for coordinating transactions and events within the overall system. Just exactly where the host application sits will depend upon the nature of the application being considered. On a stand-alone workstation using biometric verification for log on purposes, for example, the host application will naturally be resident on the workstation. Within a broader networked application, the host application may reside on an application or web server somewhere within the overall architecture. Indeed, the host application itself may be multitiered within a client server infrastructure model. Yet another variation may lie within a proprietary network of biometric devices, where the network master is in the form of a personal computer running the appropriate software and the biometric devices represent network nodes. In such a situation, the network master is effectively the host application.

The host application symbol will be used often as we describe the overall functionality of the system and how data is managed in relation to the various transactions and administration functions within our application. In this respect, there will undoubtedly be interfaces with other entities, such as back end data sources, which will need to be defined and described, and so the host application symbol will appear in various BANTAM maps as appropriate. In many applications, the host application will warrant a BANTAM map of its own in order to define the various system platform and communication requirements; whether it is supplied with, or compliant with, an existing API; which external data sources it is compatible with and what method of accessing those data sources is supplied; what the various interfaces are; and so on. This will also be the case where the biometric or related functionality being defined is to be absorbed into a larger existing application.

The user interface symbol

The user interface symbol may be used to depict a variety of user interfaces within a typical application. These may range from a simple LCD panel integral to a biometric device to the software interface for a network based system. In addition, there are different elements of systems functionality to consider, from the operational user interface to administration functions such as enrollment or report generation from the host system.

Typically, the user interface symbol will be used to indicate points within a process where messages are sent to the user interface to communicate a transaction result or other operational condition. It will also be used to indicate points where input is required from the user via such an interface.

The frequency of use of this particular symbol will depend very much upon the application in question. It might be minimal where the system is essentially of stand-alone configuration with a single point of presence. However, with a more complex transactional network based system one would expect this symbol to be used extensively. In such a case, the symbol may be used with appended attributes, for example, to indicate precisely within which screen the messaging or user response is taking place. It may even be pertinent to produce a separate BANTAM map for the user interface in order to list the available messages and potential user responses. This may be particularly useful in situations where one is anticipating a coordinated user operating environment, such as a kiosk or physical access control booth. It may be additionally useful to describe the administration functions and operation where this functionality is integrated into either the biometric device itself or a composite kiosk. The user interface symbol is thus a versatile component within the BANTAM tool kit.

The external component symbol

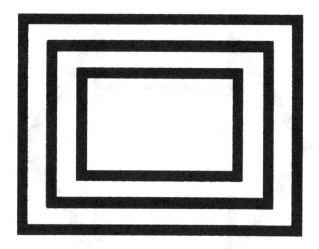

The external component symbol may be used to indicate the presence of an external component within the biometric infrastructure. This may be used, for example, where a biometric verification transaction results in a message being sent to another application, or where another device is activated in some manner as a result of the biometric verification event. Hence, the external component may be either a software or hardware component.

It may also be used in instances where the external component calls the biometric verification service. This may occur, for example, where additional authorization or authentication is required in order to access a software function or perhaps specific data within an application. The external component in this case will be the core software application which will call the biometric verification service and receive back a verification result accordingly. In such a case, there may be additional external components such as directories or application-specific access control lists which have relevance to the biometric process.

When the external component is a hardware component, such as a turnstile for example, then we may also describe its attributes, such as operational states, visual indicators and so on. This will be particularly pertinent within Logical Scenario Maps and Functional Scenario Maps. The external component symbol may also be used in the architectural context to indicate the presence of required systems infrastructure components, such as web servers, application servers, client devices *etc.*, where these have a direct influence on the overall biometric process or where there is a particular relationship which needs to be understood. The external component symbol is therefore a somewhat flexible, but very important part of the BANTAM tool set which is likely to appear frequently within the BANTAM maps.

The interface symbol

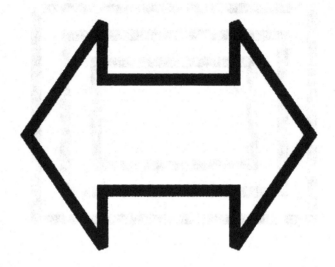

The interface symbol has a very specific use within the BANTAM methodology. It is used both to depict the presence of an interface, and to describe precisely what that interface is. Such an interface may be manifest either in software or hardware according to the situation being described. For example, there may exist a physical interface component to connect a biometric capture device to a host PC via its RS232 serial ports, or perhaps to convert a proprietary network node output to RS232 or USB for final connection to the host PC.

Alternatively, the interface may be a software interface used to access data on another system, such as a directory for example, where we may qualify the interface as being LDAP compliant. Or perhaps we are writing transactional data to an existing database via an ODBC interface. In cases such as these, we may further qualify the interface by adding attributes as required in order to specify our precise requirements, whether they be around a particular database schema or some other criteria which must be understood. A complex interface may indeed have a BANTAM map of its own in order to specify an appropriate level of detail for application developers and systems integrators. Capturing and documenting this type of detailed information is good practice and can save considerable time after the event when it is necessary to update or change one of the associated components. This is one of the key benefits of the BANTAM methodology; being able to articulate the system requirements consistently and clearly in a manner which enables engineers and support personnel to subsequently examine the documentation and understand exactly what is going on.

The database symbol

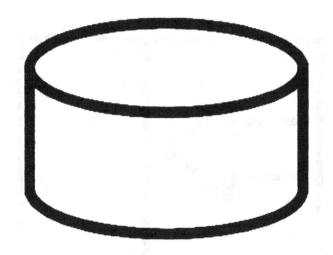

The database symbol is, naturally, used to depict the presence of a database within the BANTAM maps. The level of detail associated with the database symbol may depend upon the map in question and precisely what it is we are trying to describe. The database may not be a conventional database, but a specific storage sector within a biometric application where reference templates are stored. When an identity verification transaction is undertaken, the reference template will be called from the database and used by the matching engine in order to return a match or no match result. The database may take a variety of physical forms, from a simple storage area on a chip card, to a dedicated memory area within a biometric device, or a software related database within the host. We would therefore use the database symbol to illustrate this within our logical and functional maps as appropriate, adding attributes in order to clarify relationships where appropriate.

Alternatively, we may be using the database symbol to illustrate the presence of a transactional read/write database. Such a database may be used to log biometric verification transactions, either in isolation as part of the biometric system, or perhaps as part of a larger existing database logging higher level transactions which happen to include the biometric verification details. In this instance, we shall almost certainly wish to append attributes to the database symbol in order to describe the schema, or at least the relevant parts of it. If this is complex in nature, we might produce a separate BANTAM map for the database and, in the case of an existing database, we would certainly wish to describe the pertinent relationships within an Object Association Map. This is an area where the use of BANTAM could save considerable time in translating operational requirements into deliverable systems.

The directory symbol

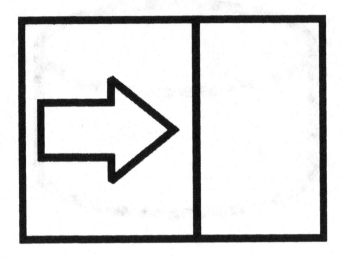

The directory symbol is used to depict a specific database which may be used in conjunction with our application, either directly for storing biometric and token data, or indirectly for accessing other user related data already stored within the organizational IT infrastructure. In this respect, we should not wish to duplicate data which already exists, both in the interest of elegant systems design and also with regard to the maintenance issues around maintaining overlapping databases.

Directories are a special kind of database which have been optimized for speed of access and concurrency around user related data. Most organizations use some sort of directory service to manage access to their computer networks and applications, in conjunction with an authentication engine. A typical directory will include all relevant information about users (usually employees) as well as systems related objects. It may or may not be possible, or desirable, to store individual biometric templates within the core directory also depending on the precise nature of our application. However, it is likely that, where a directory exists, our application may wish to access it in order to obtain user related data for reporting purposes. In such circumstances, it is important to distinguish between a database provided by the biometric or token system, and the core organizational directory. The directory symbol has therefore been provided for that purpose. It may be used at various levels within the BANTAM documentation, and will often have appended attributes to indicate precisely which data is being accessed. Within the more detailed maps, such as Object Association Maps, further detail will often be required in order to describe the relevant database schema and any particular protocols which need to be understood for transactional purposes.

The engine match symbol

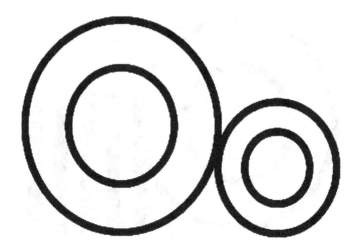

The engine match symbol is used to indicate the presence of the template matching or authentication engine within the BANTAM maps. The authentication engine is the software component which undertakes the matching process between the live biometric sample and the stored biometric reference template. It will typically be a proprietary component and it may be integrated into a suite offering additional functionality. However, it will often be useful to refer to it distinctly within our BANTAM documentation, especially within Systems Architecture Maps, Logical Scenario Maps and Functional Scenario Maps. In such cases we shall often wish to show precisely when and where the matching process is being undertaken. In this respect, the engine match symbol will typically be used in conjunction with other symbols, such as the database symbol, or perhaps token and token reader symbols, in instances where the biometric reference template is stored upon a portable token (such as a chip card).

When vendors are using the BANTAM methodology to respond to requests for information, they may wish to append attributes to the engine symbol in order to describe its typical performance. Indeed, they may wish to create a separate map in order to qualify performance parameters with recommended systems criteria, such as processor speed *etc*. They may also wish to provide additional information about the matching process and exactly how it is working, in order to place the quoted performance in context, as well as depicting the architecture around this process.

The engine match symbol is an important element within the BANTAM notation as it represents the heart of the authentication process. Understanding exactly how and where this process takes place is fundamental to understanding the broader systems design issues within any proposed application.

The engine create symbol

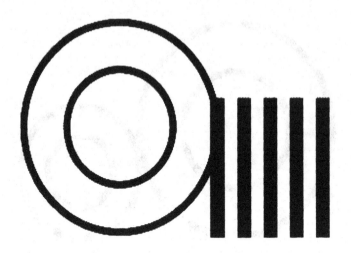

The engine create symbol is deliberately distinct from the engine match symbol described previously. While the engine match symbol is concerned with the template matching or authentication process, the engine create symbol is primarily concerned with the creation of the reference biometric template. From a systems architecture perspective, the two processes may or may not be undertaken in the same place or by the same component. It will therefore be useful to separate this functionality within the BANTAM maps where applicable.

An example of where the two processes might be separate could be when the reference templates are created within the core host system, but the actual matching process is undertaken at remote clients, or perhaps even on a portable device such as a chip card, using an entirely separate engine component. In such a case it will be important to describe this effectively within the application design documents and show what is happening at each point in the process. It may also be that a particular system can match templates which conform to a particular standard, but have been created within a completely separate application, administered by an equally separate administration. The engine create symbol is provided in order to facilitate the description of these cross-functional requirements.

The engine create symbol will probably be used in several of the BANTAM maps, as template creation will be discussed at various levels, from the conceptual to the fine detail of systems design and implementation. The symbol will also often be used in association with others, such as the database, directory, template, token and others as required to describe the various processes around biometric template creation. It will also often be qualified with the addition of attributes and other information as necessary to describe its full functionality.

The process symbol

The process symbol is a particularly versatile element within the BANTAM notation. It is used to depict an instance of a particular process which, in itself, may be described in greater detail elsewhere. In addition, there are various levels at which a process may be referred to. For example, when describing lower level software operation, a process may represent a call to a particular function, such as 'get data' or 'write transaction'. At a higher level, it may describe an overall systems related process such as 'capture sample' or 'produce reference template'.

The process symbol may also be used at a much higher level to represent a set of user related or administration actions such as 'enroll user' or 'produce report'. In essence, the process symbol encapsulates a function and/or action, or a set of functions and/or actions, in order to represent them as a single, logical process. Naturally, this process may be elaborated if desired within a separate BANTAM map in instances where it is important to understand the detail of such a process and how it relates to other system components or, indeed, other processes. This concept illustrates the flexibility, portability and power of the BANTAM methodology. Higher level maps such as Application Logic Maps or Logical Scenario Maps may use the process symbol extensively, while lower level maps may describe the processes in sufficient detail for application developers and systems integrators to actually develop the application. In between, business analysts and systems analysts will add the necessary definition to each process. The interesting point being that the process is described using a common language and common identifiers across the various project factors, ensuring a common understanding of requirements and functionality. This in turn dramatically reduces the likelihood of misunderstanding or ambiguity in defining and articulating the processes which form the practical application from the end user's perspective.

The reference symbol

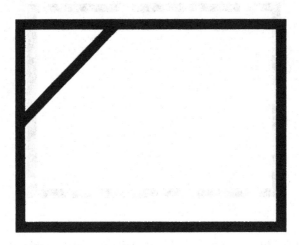

The reference symbol has some synergy with the process symbol in that it may refer to a superset of information held outside of the immediate diagram. The primary difference in this respect is that the reference symbol explicitly refers to another document. This document will almost always be another BANTAM map, referenced by the appropriate document reference on the map concerned, however, it could also be a non BANTAM document which is supplied as reference material within the overall document set. For example, the reference symbol may point towards a manufacturer's product brochure or specification sheet. Alternatively, it may point towards an in-house research paper which offers an in-depth explanation as to why a particular process is important. In fact, the reference symbol may point towards any document which has a unique identifier and is relevant to the situation being described within the host BANTAM map.

The reference symbol will often be used in association with other symbols to indicate that a further level of information is available. For example, we may find the reference symbol adjacent to a biometric device symbol, pointing towards a specification sheet or some further explanatory information. We may also find it adjacent to a process symbol, pointing towards a more detailed explanation of the process being referred to.

It is important, however, that the reference symbol be used sparingly and only where the map author feels that the additional detail provided in the reference is essential to understanding the host document. Ordinarily, one would not expect to find more than a handful of reference symbols within an entire BANTAM documentation set.

The object class symbol

The object class symbol is a unique entity within the BANTAM symbol notation suite. It refers not so much to the presence of a physical object, or a process, but to the attributes and methods within an object and their relationships with other objects. This will primarily be of interest to application developers using BANTAM Object Association Maps.

The object class symbol may be used in a variety of ways according to the context of the BANTAM map and exactly what we are trying to depict. For example, the object class symbol may simply be placed adjacent to the object symbol with which it is associated in order to describe the pertinent attributes of that object. A token object for example, may have attributes of type, capacity, format, protocol, application ID and so on. Similarly, a template object may have attributes of size, format and other parameters. Other BANTAM objects will have their own attributes as appropriate and the object class symbol may be used to show them accordingly.

It is more usual, however, that the object class symbol will be used within the BANTAM Object Association Maps. In these cases, object class symbols will show not only the attributes of key components, but exactly how they relate to each other. This is particularly important when designing transactional databases, or configuring the biometric application for integration with existing enterprise directory services. The experienced developer may liken these to entity relationship diagrams, although the BANTAM notation and methodology allows us to be more specific about the objects and associated functionality being described.

The result positive symbol

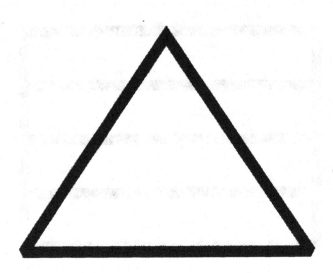

The result positive symbol is, as might be supposed, used to indicate the return of a positive match from the biometric matching engine. It will ordinarily be used in process related diagrams in order to show the logical flow of events in relation to a biometric verification transaction. For example, when a live captured template is compared to the reference template for the claimed identity, the biometric matching process will return a positive or negative result according to a certain threshold or measure of similarity. What happens next within the application will depend upon this result. A positive result will usually grant to the user the benefit being sought on the understanding that their true identity is as claimed and that they are therefore eligible for the benefit in question, whether that be access to a computer, a specific network domain, physical access through a portal, an ATM transaction or some other benefit as defined by the application.

In addition, a returned positive result would ordinarily trigger a series of system related events which, in turn, will be depicted within the BANTAM documentation. These might include the triggering of a particular dialog within the user interface, the initiation of a record to be logged within the transaction database, the referencing of other data within a separate database or directory, the resetting of biometric device status, and other events as appropriate to the application in question. The result positive symbol will therefore be used quite extensively within the BANTAM documentation in order to show clearly what happens as a result of a positive template match. It will also be used at different levels of granularity to show both conceptual logic at a high level and actual data flows within the more detailed BANTAM maps.

The result negative symbol

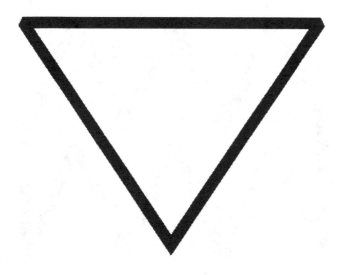

The result negative symbol is used in the same way as the result positive symbol, in that it indicates the returned result from a biometric template matching instance. It will ordinarily be used in process related diagrams in order to show the logical flow of events in relation to a biometric verification transaction. For example, when a live captured template is compared to the reference template for the claimed identity, the biometric matching process will return a positive or negative result according to a certain threshold or measure of similarity.

A returned negative result will initiate a series of events according to how the application is designed and constructed. Primary among these events will be denying access to the benefit sought and communicating the result to the user. In many instances the user will be offered another chance of verification, in which case the entire biometric capture and verification process will be undertaken again. This is important, as there may have been a good reason for the negative result being returned, caused by conditions outside the immediate control of the user, for example, momentary system or communication failure, capture device malfunction and so on. The communication sent back to the user interface is thus important and may take several forms, according to the exact condition. The result negative symbol will therefore often be used in association with the user interface symbol and qualifying comments as appropriate. Both result negative and result positive symbols will also be used frequently in conjunction with the database, engine and template symbols in order to show the logical functions around a biometric verification transaction.

The transaction symbol

The transaction symbol is used to illustrate a transaction instance. Typically, this will be a biometric verification transaction, taking place within a broader overall process, but it is not necessarily confined to this example. Indeed, the transaction may be a simple data exchange transaction between different elements of the same application, or perhaps between different applications. For example, the host application may wish to reference data stored in an external database and may need to pass certain criteria accordingly. In such a case, the transaction symbol may be used in association with other symbols to indicate precisely how this is achieved within the overall systems and functional architecture.

Within the context of a biometric transaction, the symbol may be used to illustrate the precise point of the transaction initiation within the overall process framework. It may also be used to indicate where, as a result of biometric verification, a related transaction is initiated between applications. This might occur, for example, where the biometric verification process is being conducted on a local computer, after which communication is initiated with a remote application stored elsewhere on a local network, or perhaps a web server via the Internet.

The transaction symbol will generally be qualified with various attributes appertaining to its use within a given instance. It may additionally be used in conjunction with the external component, interface and other symbols in order to describe a particular transaction more fully in architectural terms. Furthermore, it may be used in context within most of the BANTAM maps.

The action symbol

The action symbol, as its name suggests, is used to depict a specific action within the BANTAM maps. It will appear most frequently within the higher level BANTAM maps such as the Application Logic Map and Logical Scenario Map, although it may be used anywhere if deemed applicable to the project at hand. An action may be a user related action, such as the user initiating a biometric verification transaction or offering a token to a token reader for example. It may be an administrator action, as required, for example, to query the system for information, or perhaps to initiate an attended enrollment. An action may also be a systems related action in response to a certain condition or set of conditions. For example, the system may initiate an action to operate a physical barrier, or perhaps to seize a token which has passed its expiration date or which is believed to have been used fraudulently. In short, the action symbol may be used to highlight the occurrence of a specific action within a set of processes, whether human or systems related.

Ordinarily, the action itself may be self-evident or easily described within a short appended statement. However, in cases where the action is deemed unlikely, or of particularly high consequence, then a detailed description should be provided within the Map Explanatory Notes. If the action may lead to a complex chain of events, then a separate Miscellaneous Definition Map may be used and referenced from the primary BANTAM map with the reference symbol.

The capture symbol

The capture symbol is used to depict the instance of a biometric sample capture. This may occur either during enrollment or subsequent identity verification attempts and it will need to be adequately described within our BANTAM documentation according to the process being considered.

This symbol will be used in a variety of maps and at a variety of levels. It may appear frequently within Logical Scenario Maps and Application Logic Maps, for example, to simply illustrate a sample capture instance. It may also be used within Functional Scenario Maps in order to depict the precise point of sample capture within a broader overall function, and hence its relationship to data flow between components and throughout the system. In this respect, it will often be used in association with the template, database, directory and engine symbols to provide clarity around a specific operation.

The capture symbol may be used with appended attributes in instances where we need to describe the data being captured and how this is to be utilized between components, which may or may not have been sourced from the same supplier. This may also be important for developers wishing to integrate a particular sensor device into another component which, in turn, will be communicating with other system elements. The capture symbol thus represents a good example of the flexibility and portability of the BANTAM notation.

The authenticate symbol

The authenticate symbol is used to indicate the authentication process. In biometric terms this will mostly relate to the template matching process, but the symbol may be used in a variety of contexts where biometrics may or may not play a part. For example, we may be describing a token only application, where authentication effectively means authenticating token data against a reference database. We may be using the symbol to describe authentication deeper within an application where directory services are used to authenticate against user names or passwords provided by the system in a process which is invisible to the user. This may occur, for example, where we have implemented a single sign on approach which uses strong authentication at the start of a session, followed by a faster, automated weak authentication within the session, using password data generated or released by the previous stronger authentication.

The authenticate symbol may also be used at various levels within the BANTAM documentation, from the higher level logical maps, to the functional and architectural depictions of the various system scenarios, indeed, whether or not biometric or token technology is actually deployed within the part of the system being described. It is thus a flexible, and very important symbol within the overall BANTAM notation and it will often be used in association with other symbols in order to illustrate an overall authentication process. Typically, the precise details of the authentication process indicated by the symbol will be described within the Map Explanatory Notes for the BANTAM map in question.

In conclusion

The BANTAM symbol set has been designed to be simple and intuitive in use. The overall symbol collection has been deliberately kept small in number (22) to avoid both confusion and also the need for a steep learning curve. The symbols may be used with or without descriptive titles and are provided in both forms accordingly, although it is suggested that titled symbols be used in instances where BANTAM map documents will be shared among different groups within a given project, or used in conjunction with RFI and RFP documents.

When creating BANTAM maps with the notation symbols, connections between the symbols are depicted simply as either plain lines, or lines with arrow heads to depict a direction of flow, as illustrated below.

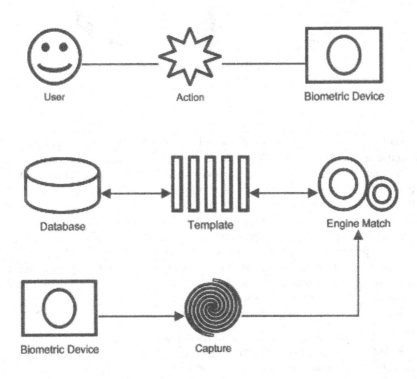

Figure 3.1 Connecting BANTAM symbols

In keeping with the BANTAM philosophy, this is deliberately kept simple and intuitive with no complications or variances according to the type of map used or other conditions. Arrowheads may be used to indicate either a single direction of flow or bi-directional flow where appropriate and will take the form of a solid arrowhead graphic appended to the connecting line. Most popular graphics programs will include the facility to add such connecting lines as standard, as well as the ability to import the BANTAM symbols easily, and should therefore be capable of producing BANTAM diagrams, which in turn may be pasted into the standard BANTAM document templates, provided in Microsoft Word format.

3.3 BANTAM documentation

The BANTAM methodology provides for a full set of document templates in order to promote consistency and ease of use. These templates are included on the accompanying CD-ROM and include the following:

- Application Logic Map

- Logical Scenario Map

- Functional Scenario Map

- Systems Architecture Map

- Object Association Map

- Miscellaneous Definition Map

- Map Explanatory Notes

- Request for Information

- Request for Proposal

- Training Schedule

- Business Case

Together with the BANTAM symbol notation, these documents allow for the definition and description of virtually any element within a typical biometric, token technology, or related project, right through to procurement, implementation and beyond. The strength of BANTAM lies partly in its simplicity and intuitiveness, and partly in its flexibility and portability over every phase of a typical project. This applies even after implementation, when the BANTAM document set may be used for training and maintenance purposes as well as forming a valuable program archive.

The overall concept of the document set is that it should be capable of capturing as much or as little detail as required at each phase, while remaining simple and streamlined. The primary maps are therefore restricted to just five in number, plus a Miscellaneous Definition Map to capture any unusual circumstances or configurations within a given project. Each map is accompanied by a Map Explanatory Notes document to provide a textual description of the process at hand, plus specific details of system components where appropriate, thus providing an intuitive graphical representation, together with the opportunity to add relevant detail for each element or process being described. In addition, the Request for Information (RFI) and Request for Proposal (RFP) maps extend the BANTAM concept right into procurement, where the methodology may also be used for evaluating proposals against a clear reference and in a like-for-like manner.

Each BANTAM map document has a header which gives the document type followed by the primary document reference details as follows:

- Document title – the name of this particular document.

- Document ref. – the specific reference for this document.

- Prepared by – the individual who created the document.

- Date – the date on which the document was formally issued.

- Project name – the name of the applicable master project.

- Type – the type of document, working or final.

Baŋtam | Application Logic Map

Document title	Example ALM	Document ref.	ALM 0001		
Prepared by	Julian Ashbourn	Date	14/07/2001		
Project name	Network Access	Type	Working	X	Final

Figure 3.2 BANTAM document header

Document references should consist of the map type initials, for example, ALM = Application Logic Map and FSM = Functional Scenario Map, followed by the internal document reference number on an incremental basis. This will help with document management, especially where multiple projects are being undertaken simultaneously. The document type identifier is used to distinguish between a working document, which may be subject to revision, and a final document, which is agreed by all concerned and forms part of the official program document archive. Only final documents may be used for subsequent training and maintenance purposes. The process under which a document is declared final will no doubt vary slightly from organization to organization, but it will typically be under the control of the program manager, who will define the process for submitting and agreeing such documents. This may take place within a working or steering group meeting for example, where representatives from all relevant departments are present. It is outside the scope of BANTAM to prescribe such a process, but it is highly recommended that organizations using BANTAM define and document the means by which documents may become final versions. Footers are provided within the BANTAM document templates in order that organizations may add their name and copyright to each document.

It is important that the standard BANTAM document templates are used and not altered in any way. This provides ongoing consistency and ensures that documents are readily understood by those utilizing the BANTAM methodology. Similarly, the standard BANTAM symbol library should be used at all times without modification of any kind. Again, this ensures that the BANTAM documentation that you produce will be readily understood, not only throughout your own organization, but among third party contractors and suppliers who may have an involvement in your particular project. This standard documentation and notation set has been designed to cover virtually all eventualities within a typical project. Any unusual processes or application specific situations which do not fall within the scope of the primary BANTAM documents, may be easily covered within the Miscellaneous Definition Maps and the accompanying Map Explanatory Notes. The various BANTAM documents are described below.

Application Logic Map

The Application Logic Map (ALM) is a high level map which may be used to describe the basic concept of the proposed application in relatively simple terms. It will usually be produced by the end user project sponsor, probably in association with a business analyst or consultant who will understand the working processes involved. The primary purpose of the ALM is to describe the requirement, as understood by the end user, using the BANTAM symbol notation to show the logical elements of the proposed application and how they might fit together. The accompanying Map Explanatory Notes (MEN) will provide a textual description of the overall requirement, together with details of known system components as well as specific components under consideration for this project.

The initial version of the ALM may be used as a discussion document between the various internal and external departments involved, with the program or project manager noting any required changes accordingly. Ultimately, a final version of the ALM will be produced which sets out the high level application logic as agreed by the project team. This finalized high level view of the application will serve as a reference guide, or route map, for the more detailed maps which describe the major processes and systems elements. The ALM will, in the vast majority of cases, be included with Requests for Information (RFI) and Requests for Proposals (RFP) in order that prospective suppliers understand the requirement in the same context as the end user. It is, in essence, the starting point in the chain of documentation which fully describes the project under consideration and for which BANTAM has been designed.

Logical Scenario Map

In many cases, a biometric or related application will be quite complex and may interface with several other systems. Furthermore, the practical implementation of such an application may cut across several operational departments within a given organization, and maybe across several organizations. In such circumstances, it is often useful to break down the larger application into logical sections which may be defined and designed as holistic units, providing the necessary focus and clarity to best develop each section. The BANTAM methodology allows us to do just this with the provision of Logical and Functional Scenario Maps. In many projects, there will be multiples of these maps in order to accommodate the different sections as described, while sometimes, for smaller projects, there may be just one or two.

The Logical Scenario Map is the starting point as it defines and describes the process being undertaken in this particular sector of the total application. The map will often be produced in association with business and systems analysts, who will understand both the existing process and the proposed new process. Indeed, it may be worthwhile in many cases to document both the old and new in order to share this understanding among those involved in developing the application. This would also be good program management practice in providing an audit trail of how a given process was developed. The Logical Scenario Map will make full use of the BANTAM notation in its graphical description, although it will probably not be necessary to add too many attributes or references at this stage. The accompanying Map Explanatory Notes however will probably be very rich in their detailed description of the process at hand. In many cases, the high level Application Logic Map may reference several Logical Scenario Maps in order to build a detailed picture of the application incrementally from a logical or operational perspective. The Logical Scenario Map is therefore a key component of the overall BANTAM methodology.

Functional Scenario Map

Whereas the Logical Scenario Map describes a particular process from a conceptual or operational point of view, the Functional Scenario Map gets right down to the detail of exactly how a given process will work from a technical perspective. This may constitute a fairly complex document in some cases, especially if there are several interfaces between components relative to the process or function being described. Fortunately, BANTAM makes life simple in this respect, by allowing you to break out complexities into separate documents and use the reference symbol to point towards them. In addition, the Map Explanatory Notes allow you to describe any unusual circumstances in as much detail as is required.

The Functional Scenario Map has a very important role to play within a given biometric or related project. It will often act as a bridge between the end user aspiration and the application developer's deliverable system. Having scoped out the application at a high level with the Application Logic Map and added more detail around proposed processes with the Logical Scenario Maps, the Functional Scenario Maps can now test the technical feasibility of supporting a given process and the associated system functionality. It will often act as a discussion document between the project sponsor or manager and the development team, in order to drive out the detail of how an application will be designed and built. Where the development team is a third party supplier, this is especially important as any lack of clarity and understanding at this stage could result in costly delays or, worse, a delivered system that simply doesn't meet the requirement.

In a typical project, there may be several Functional Scenario Maps, in order to describe each major part of the application adequately. Furthermore, some of these maps may reference other maps where required in order to describe a complex technical operation or interface fully, to the extent necessary for the development team to understand the requirement fully and be able to build the system accordingly. Once the various Functional Scenario Maps have matured into the final document versions, they will effectively form the blueprint of the application from a development perspective. They are therefore a vital element of the BANTAM methodology.

Systems Architecture Map

The Systems Architecture Map has a slightly different purpose, compared to the previously mentioned BANTAM documents, in that it is not primarily concerned with the new biometric or token technology application, but with the overall systems infrastructure and therefore the compatibility of the proposed new application with the existing architecture. There are various ways in which the program manager may wish to use Systems Architecture Maps. In all cases, it will be useful to describe the current architecture, including network topology and bandwidth, protocols and systems interfaces. The Systems Architecture Map will facilitate this, together with the BANTAM notation, using component attributes and the reference symbol where necessary in order to break out a particularly complex area onto a separate map.

Having accurately described the existing architecture, it will then be useful to produce a second map, showing exactly how the proposed application will integrate into this infrastructure. In this respect, it may be necessary to break out onto a separate map any areas of the existing architecture which need upgrading or otherwise altering in order to accommodate the new application. The BANTAM reference symbol may be used in such a case to point towards the detail map. In any event, it will be important to show the detail of any interfaces and attendant protocols where new system components are to be integrated into the wider systems architecture perspective.

The extent to which the Systems Architecture Map is used within your project may depend on how well the existing architecture is currently documented. One of the strengths of BANTAM is portability and the opportunity for reuse of the BANTAM documentation. The Systems Architecture Map is a very good example of this. The user will find such maps invaluable, not just for the new biometric or token technology application which produced them, but across a broad range of scenarios where it is useful to have such detailed and logical information available. Indeed, in many instances, it would be well worth producing a set of BANTAM Systems Architecture Maps for your organization, whether or not you are currently contemplating a new project. A clearly defined and documented archive of your overall systems architecture should be a prerequisite of any professional organization, no matter how large or small. If, on the other hand, you already have a very well documented architecture, then you may choose to use the Systems Architecture Map purely to describe the relevant parts of the architecture which will be affected by the new application.

The other major benefit of this map is its use as a discussion document with potential suppliers, ensuring that they fully understand your architectural situation and how you anticipate the new application being absorbed into it. Furthermore, they may then use the same BANTAM methodology in order to respond to such issues within their proposal, making it easy for you to understand their suggestions and technical propositions. Lastly, the Systems Architecture Map will form a valuable part of your program archive, describing exactly what the architecture was at the time of implementation, and what changes, if any, were made as a result of this project. This sort of information will be invaluable to maintenance and support staff long after the event.

Object Association Map

This map will be of particular interest to application developers. Its purpose, as the name suggests, is to depict the associations between objects accurately. These objects may represent any system components, whether they be manifested in software or hardware. From a software perspective, the Object Association Map will frequently be used with respect to databases and the management of data between entities in the instance of a transaction. For example, the user may enter a reference, either by a direct entry on a keypad or a card swipe which, in turn, references a database in order to associate this reference with an individual name and perhaps retrieve a specific biometric template. This information may or may not reside in the same place. For example, you may be using an organizational directory for the primary personnel information, but storing the biometric templates elsewhere. After a matching process has been undertaken, a record of the transaction may be stored in quite another database. In such a situation, we will need to understand the relationships between these various data sources and exactly how they will communicate with each other. The Object Association Map will facilitate just this, providing a convenient reference for developers, whether they be internal or third party personnel. As with all the BANTAM maps, its development may be iterative, being a working document and subject to change while the application is being scoped and designed, and subsequently being signed off as a final document for development purposes. It will therefore be an extremely useful discussion document throughout the various project development phases. In addition, Object Association Maps in their final form, will constitute a valuable part of the overall program archive and will prove extremely useful within related developments or enhancements to the main project at a later date.

From a hardware perspective, the Object Association Map may also have a role to play, especially where different hardware components need to be integrated within a particular process. For example, we may be using chip cards and biometrics within a PKI environment and need to understand exactly how and what data is being used to facilitate the release of the private key. There may be standard or nonstandard protocols being used, or perhaps translation layers between components, and other details which need to be defined and understood. Furthermore, the relationship of these components to primary databases will also need to be described, and the Object Association Map will fulfill this requirement.

Miscellaneous Definition Map

The BANTAM methodology provides a succinct documentation set which will, in most instances, prove more than adequate to cover all eventualities associated with a typical biometric or token technology application. However, this documentation set has deliberately been kept small in number in order to promote ease of use, and it may be that a particular situation exists within your project, which doesn't fit comfortably within any of the standard BANTAM maps and yet is important to define and document within the context of the project overall. Enter the BANTAM Miscellaneous Definition Map. The Miscellaneous Definition Map may be used to describe any element of your project, using the standard BANTAM notation. This means that you can use all the same techniques and BANTAM symbols, but in a relatively freeform manner. Naturally, the Miscellaneous Definition Maps may be referenced from other BANTAM maps as appropriate, and *vice versa*.

You may find the Miscellaneous Definition Map particularly useful for describing related processes which, while not a core part of the application, nevertheless are important to include in overall consideration terms. You may also find this map useful for breaking out and describing a particular technical detail which needs to be understood within the broader context. In short, the

Miscellaneous Definition Map is whatever you want it to be within the context of the overall program and may be used accordingly. It adds a useful degree of flexibility to the BANTAM methodology, without deviating from the underlying principles.

Map Explanatory Notes

Every BANTAM map, such as described above, should be accompanied by a Map Explanatory Notes document. While the BANTAM notation is extremely intuitive in use, in many instances it will be useful, if not necessary, to add a plain text description of the process or technical scenario being described within the map.

The Map Explanatory Notes document facilitates this in a structured manner and ensures that any BANTAM map may be easily understood by any of the project team, regardless of their individual familiarity with the situation being described. In addition, the use of this document provides valuable information for potential suppliers, whether they be consultants or systems integrators, to help them fully understand your requirements and how they might meet them. The Map Explanatory Notes document is divided into three logical sections. Firstly, the Map Overview section provides for a plain language textual description of the process or situation being described within the host map, in as much or as little detail as required. The Component/Device Details section provides an opportunity to specify pertinent system components, either generically by technical capability, or specifically by manufacturer and model if this is important within the overall application design. Lastly, the Relative Information section provides the opportunity to refer to situations, technical or operational, which may have a bearing on the situation being described within the host map. The first overview section is mandatory, while the second and third sections may be completed at the discretion of the individual preparing the host map, according to the precise nature of what is being described and just how much detail is considered necessary. However, it is anticipated that the majority of Map Explanatory Notes documents will be fully completed in order to describe the situation accurately and fully. Once again, the use of this document adds valuable extra detail to your program archives.

Request for Information

One of the perennial difficulties with projects which utilize specialist technology such as advanced tokens and biometrics, lies in the definition of requirements in a manner which may be easily understood by all parties concerned, and the efficient procurement of the necessary technology and services which will lead to an accurate implementation of the original requirement. The standard BANTAM methodology covers the first point admirably but, in fact, also extends to the second. This is another unique aspect of BANTAM, in that it follows right through to procurement and beyond, helping you towards a successful conclusion to your original aspiration.

The Request for Information (RFI) document has been carefully designed to ensure that you can obtain information from prospective suppliers in a meaningful and consistent manner, while providing for a logical means of evaluating the responses received. In addition, vendors and suppliers using the BANTAM methodology will additionally benefit by seeing a consistency in requests which, in turn, helps them to build an increasingly pertinent view of real world requirements to enhance their own knowledge base as they move forward and develop their capabilities accordingly. The document adheres to the BANTAM principle of simplicity, but is in fact quite powerful in its descriptive capabilities, while providing a much needed standard for RFIs. The first section provides the necessary detail of precisely who is requesting information, and by when they would like a response. It also categorizes the product or service being inquired about. The second section lists all of the attached BANTAM maps with their title and reference details. The next three sections describe the basis for the request, the requirement as currently perceived and any special conditions which need to be understood in relation to the project at hand. The next section provides details about the organization making the request, the format in which a response is required, and the correct procedure for asking questions in relation to this RFI. Finally, the RFI is authenticated with details of the signatory and of course the signature in question. However, the BANTAM RFI document doesn't stop here. A response summary is included, requiring the respondent to complete a standard minimum set of information about their organization, together with other pertinent information which they may like to include at their discretion. It also requires them to list any BANTAM documents included with the response, as well as other non BANTAM attachments. Finally, it requests a signature and details of the signatory. This summary will prove a useful document when evaluating responses and brings a welcome degree of consistency to the process, making it easier to compare like with like. It also removes any ambiguity appertaining to the response and clearly references specific BANTAM documents for the necessary detail. The BANTAM RFI document represents a powerful aid to professional procurement and makes full use of the underlying BANTAM methodology.

Request for Proposal

Whereas the RFI document provides valuable benefits in the gathering of pertinent information and identification of potential suppliers in relation to your project, the Request for Proposal document moves to the logical next step of procurement. Having selected a list of potential suppliers for your project, it will now be necessary to elicit specific proposals from them, either for the entire application, or perhaps for certain key areas. In any event, we must supply potential suppliers with sufficiently detailed information to allow them to respond with an intelligently conceived proposal. Furthermore, we need to be able to evaluate the resulting proposals in a consistent and dispassionate manner against well defined criteria. Fortunately, the use of BANTAM has provided us with a clear and consistent descriptive mechanism with which to articulate our requirements. We now need to be equally clear about requesting and evaluating proposals, a matter with which BANTAM can also assist.

The standard BANTAM Request for Proposal is a six page document which covers all the fundamental requirements of such an exercise. It starts off, like the RFI, with full contact details for your organization and the date by which proposals should be received. Naturally, it also includes a unique reference number and a name for the project in question. The next section lists all of the BANTAM maps included with the RFP, followed by a Project Overview, which describes the project in plain language terms and includes a footnote listing related map references. The next page is devoted to a detailed requirements section, again in plain language terms and including a footnote of related BANTAM maps. The next section lists any special operational or architectural conditions which need to be taken into consideration, again followed by a footnote referencing the relevant maps. We then come to a section covering site visits, which sets out the relevant contacts and conditions, together with a plain language description of any related requirements, for example, in some military situations security clearance may be required, or in hazardous situations it may be necessary to wear protective clothing, and so on. A further section allows you to describe your organization. The following three sections are very important from the respondent's perspective as they clearly define the format of response required, the correct procedure for asking questions and how responses will be evaluated. This is followed by the authorized signatory details and the signature itself. As with the RFI, a Response Summary is included which sets out the respondents details, including the principal company officers and other pertinent information. It also lists similar projects undertaken and details of a reference site, together with a list of attached BANTAM maps and non BANTAM attachments as appropriate. The next section provides for an estimated cost summary, broken down into equipment, installation and other costs, and finally the authorized signatory and signature from the responding organization.

When you receive back the completed BANTAM RFP documents, you will have a consistent set of proposals which may be easily evaluated on a like-for-like basis, without being unduly swayed by proprietary documentation or marketing speak. BANTAM is neutral and portable across organizations on both sides of the procurement fence. This benefits all concerned by providing a level playing field as far as project proposals are concerned. As for the politics of such a situation – well, that is another matter but, again, BANTAM will be providing a much needed clarity and consistency where it really matters, in the technical and operational detail of what is being proposed.

Training Schedule

Training plays a vital part in any application or operational system, and this is especially so when technologies such as biometrics are employed. BANTAM allows you to configure and organize your training requirements efficiently via the Training Schedule. This document identifies a particular schedule and its purpose, and lists the BANTAM maps which are referenced as part of the schedule. It also identifies the target audience and any specific hardware or software required in order to undertake the training. A schedule overview describes the schedule and its objectives in plain language, and any certification associated with the schedule is listed accordingly.

The Training Schedule contains a number of independent training modules which, together, make up the overall schedule. These modules are listed in a summary section, together with their reference identifiers. A separate, but attached document is then created for each training module and contains the tutorial material itself which, in turn, will reference the appropriate BANTAM maps. Using the Training Schedule facilitates the creation of a structured training program for each important element within your overall program, for example, you may wish to create separate schedules for administration personnel, general users and maintenance engineers. If you decide to use the BANTAM Program Manager software, the training module within that program allows you to configure your training schedules easily and assign personnel to different schedules as appropriate. As previously indicated, training is a vitally important item within your overall program. The BANTAM Training Schedule brings clarity and consistency to this area, just as it does with program definition and systems design.

Business Case

The BANTAM Business Case document enables you to define and capture the original business case in a clear and consistent manner. It encourages you to think about the key issues for any business case, including cost/benefit analysis, other options considered, associated risks and much more. The document itself is really self-explanatory and very easy to complete and you may of course already have such a standard document within your organization, but if not, the BANTAM Business Case document will prove invaluable in bringing clarity and consistency to this key area, especially when dealing with a number of projects.

The document starts with a header outlining the project name and reference together with details of where the proposal originated and whether the document is a draft working document or a final version. It then proceeds to an executive summary which highlights the project fundamentals. These start with a proposal overview which explains the project in plain language terms, a summary of the perceived benefits provided by the proposed project, the reasons for change, or why such a project is needed in the first place, other options considered in addition to the proposal being put forward and why they were rejected and, finally, the risks associated with the project as proposed. This section provides a pertinent summary for discussion and elaboration within a project board meeting, or whatever forum you have for approving business cases within your particular organization.

The next section of the BANTAM Business Case document provides for a more detailed project description. This contains sections for the technology being proposed, where details of the technology and its suppliers may be noted, the fit with the existing infrastructure, where details may be given as to how this has been considered and catered for, key milestones and timescales, where the pertinent deliverables may be noted accordingly, a detailed cost benefit/savings analysis, where figures may be provided in support of the business case, details of how the project would be funded and, finally, estimated project costs. Now we have the detailed analysis to support the executive summary.

The final section of the Business Case document provides for a summary of associated BANTAM map documents and a provision for three signatures together with associated demographic details. We thus have a format which can pass through various iterations as a working document, before finally being signed off by those with the necessary authority to do so. This also facilitates the situation whereby a business case may be initially rejected, perhaps because more detail is required, and needs subsequently to be re-presented before final approval, or indeed, final rejection as the case may be. The BANTAM Business Case document makes the whole process straightforward and ensures that the basics are covered for any of your projects.

In conclusion

The BANTAM document set is an extremely powerful aid to program management, application development and procurement within a typical biometric, token technology or related project (although the use of BANTAM is not restricted to those areas, it may be used effectively in many scenarios). Part of the strength of BANTAM lies in its inherent simplicity and the fact that it is very easily learned and absorbed without the steep learning curve (or cost) required by some methodologies. The BANTAM document set illustrates this concept nicely, bringing clarity and consistency in a powerful and flexible package, which covers just about any eventuality connected with your project. A full set of document templates is provided on the CD-ROM included with this book, which the reader may use immediately. The only condition to the use of BANTAM is that, in the interests of consistency, the documents should not be altered in any way. Any enhancements or additions to the BANTAM documentation set will be managed centrally and posted on the BANTAM section of the Avanti web site. However, it is anticipated that this will be an infrequent occurrence as BANTAM is already very comprehensive in its coverage and should provide everything you need to manage your works program.

4. Using BANTAM

This chapter will provide a more practical introduction to using the BANTAM notation and document set. In this section you will learn more about the use of the primary BANTAM maps and hopefully, will be able to place this in the context of your own projects, whether they be concerned with biometrics and tokens, or related technologies. The reader is encouraged to produce some example BANTAM maps for themselves, in order to get a feel for using the methodology. This is easily achieved with the tools and templates available on the accompanying CD-ROM. The Serif DrawPlus 5 trial graphics package will enable you to produce the maps and then paste them into the Microsoft Word format document templates. To accomplish this you must import the BANTAM symbol set into DrawPlus 5 using the utility provided on the CD-ROM. If you prefer, you may use an alternative graphics program, in which case the BANTAM Symbol Selector (also provided) will prove extremely useful as a means of copying and pasting the symbols into your preferred program. This flexibility in approach enables BANTAM to be used by the widest cross-section of users without having to purchase expensive dedicated modeling software packages or having to learn a difficult operational routine.

4.1 Planning the model

When setting out on a new project initiative, it is a good idea to decide in advance which BANTAM documents you will be using and why. This allows you to tie in with your resource planning and assign document creation tasks to the appropriate individuals within your team as appropriate. Of course, it may be that for certain documents, you need to collaborate closely with external vendors or consultants in order to gather the necessary detail. You should therefore include the creation of the relevant BANTAM maps as milestones within your overall project plan and monitor the completion of them closely. This is an important point. If there is slippage in creating the proper documentation, then that in itself will introduce the possibility of misunderstanding around fundamental requirements and principles of operation, simply because the detail has not been defined and agreed. In other words, each and every BANTAM map should be regarded as a deliverable within your project, and therefore treated the same as any other deliverable with regard to status tracking and risk management if not completed on time. If you are using one of the established project management methodologies and software tools, then it will be relatively easy to build the BANTAM documentation into your original project plan and then baseline this accordingly. If you are not using a project management tool, then, at the very least, plot out your project against a timeline on a spreadsheet or other document, in order to have a base level plan to work and monitor against.

As for how much or how little of the BANTAM methodology you will need to use, this will very much depend upon the size and scope of the project being considered. In some cases, a subset of the complete document set may suffice in order to describe the pertinent requirements for your project. In other cases, you may need multiples of every single BANTAM document. As an absolute bare minimum requirement, perhaps for a very small stand-alone pilot project for example, the following document set is recommended.

- Application Logic Map with Map Explanatory Notes.

- Functional Scenario Map with Map Explanatory Notes.

- Request for Proposal.

In the majority of cases, however, more detail will be required and it will be necessary to use more of the BANTAM document set. In fact, the Miscellaneous Definition Map, may be the only document not used regularly among your projects, depending on their relative complexity.

You should also ensure that, from a practical perspective, you understand exactly how these documents will be created and published, both internally and externally. The first step will be to define and set up a repository for the documents where they will be safe, and from where they may be easily backed up as part of your standard data protection policy. For this, you will probably need to speak with your LAN administrator and agree on a specific network drive, accessible to all those who will be involved in the practical creation of the documents. In this respect, do not forget your procurement department, who may require access to the RFI and RFP documents at some stage, even though they may not have created them. You will then need to consider how to ensure that all those who need to use the documents (in addition to those who created them) have ready access to the latest versions at all times. You may decide to do this via a specific Intranet site from which documents may be viewed or printed, or perhaps from a specific Notes database or team room if your organization uses Lotus Notes. Or, indeed, you may have your own proven mechanism for sharing such information. In any event, be sure to consider the security and access control implications associated with this information. You may not wish the whole world (or even the whole of your organization) to know your aspirations in this context.

Having set up a workable repository and document distribution system, and entered the creation of the relevant BANTAM maps as milestones within your project plan, you will now be ready to produce the maps as you move through the various stages of your project. You may wish to finalize your use of a specific graphics program with which to create the graphics in order to maintain consistency when you paste them into the Microsoft Word document templates. The trial version of Serif DrawPlus 5, included on the accompanying CD-ROM, is one option in this respect, and represents a low cost tool which is easily capable of producing the BANTAM graphical elements. However, as the BANTAM symbols are also provided for you in both bitmap and metafile formats, you may prefer to import these symbols into the graphics application of your choice, in which case, simply follow the instructions supplied with the application in question. When you have chosen a suitable application, be sure to communicate this to all personnel who will be involved in producing the BANTAM documents and ensure that they have access to a licensed copy of the program. You may even like to produce a brief instruction sheet, advising on which tools to use, document version control, and where the documents should be stored.

With regard to document version control, BANTAM provides for a unique reference for each document as well as a document creation date. It also allows you to tag each document as either 'working' or 'final' in order to draw attention to the document status. In order to benefit fully from this convention, it will be necessary to formulate and operate your own process for managing document versions. One way of achieving this is to set up a small document management group, responsible for signing off documents as 'final' according to a preconceived set of criteria, which acknowledges the fact that some documents will be created jointly with external parties. This group would also be responsible for highlighting documents whose completion is close to, or falling behind schedule according to the project plan. The Document Management Group (DMG) may be no more than two or three individuals, picking up this duty in addition to their primary roles, but it will prove to be a useful element of your overall program management.

The remainder of this chapter will take a look at the primary BANTAM maps and offer some suggestions around their use within a typical project.

4.2 The Application Logic Map

The Application Logic Map (ALM) will be especially useful in the early stages of the project, when you are defining the overall concept and need to verify the feasibility both with your internal IT support departments and potential external suppliers. In creating the initial ALM, you should be heavily process driven and not get too bogged down with the fine detail of biometric devices or performance claims. In addition, it is assumed that in order to get to the stage of creating an initial ALM, you have already addressed the business case for your project and reached positive conclusions as to the viability of such an idea. In doing this, you will have identified the primary problems around the existing processes and quantified their effect, either in hard financial terms, or in dependent activities. You will have come to the conclusion that biometric, token and related technologies can help you in your quest for a workable solution and that you are now ready to take the next step into further definition of your proposal and engagement of third parties as necessary in order to realize the practical implementation of your idea. Fortunately, you have BANTAM as a working methodology to help you through the various stages, and a good starting point (once you have your project plan in place) is the ALM in which you will describe the high level principles of your proposal. Just exactly what you include in the map at this stage will no doubt depend on the proposed project in question, but it may be something as simple as the following.

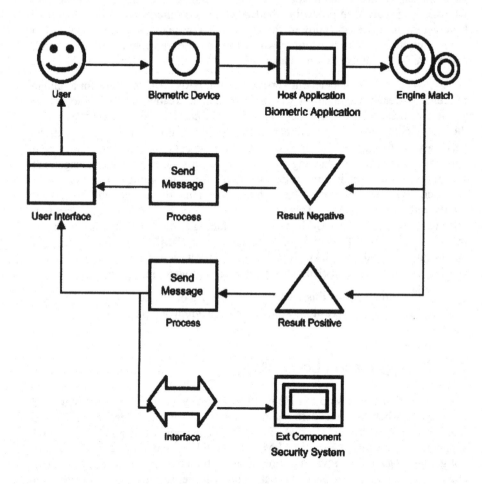

Figure 4.1 Application Logic Map graphic

In Figure 4.1 above, we depict a very simple scenario whereby the user interacts with the biometric device which, in turn, communicates with the host application, in this case the proposed biometric application. A matching process is undertaken, which yields either a positive or negative result, each of which sends a message back to the user interface accordingly. If the result is positive, however, further interaction takes place via an interface to an external component, in this case an existing security system.

Now, this information alone is not enough with which to design an intelligently conceived and optimized system, but it is sufficient to communicate the basic requirement and start discussion among the various parties in order to think through the detail. This is the whole idea of the ALM. To enable you to capture in broad terms the overall requirement and initiate intelligent discussion accordingly. Bear in mind that an ALM containing a diagram such as that depicted in Figure 4.1 will always be accompanied by a Map Explanatory Notes document (MEN) which will provide for a plain language description of the high level requirement and will be evaluated together with the ALM. You may have a crystal clear picture of the initial requirement and be able to capture it right away in such a document. More typically, the creation of the ALM will be an iterative process involving discussion with the various parties concerned and resulting in a refined and revised final ALM document, to be used throughout the project from this point onwards. Just how many iterations this takes will depend upon the complexity and clarity of the original idea and the Document Management Group will be instrumental in guiding this process through to completion of the final document. The ALM may thus be thought of as the starting point for discussing and developing your original idea, not in a haphazard fashion, but within a structured methodology which leads forward into the necessary detail required to develop the system according to your original requirements. The ALM will therefore be referred to often throughout the project and may include references to other BANTAM documents in order to expand on particular areas of operation.

4.3 The Systems Architecture Map

The Systems Architecture Map (SAM) is a key component within the BANTAM methodology and will additionally form an important part of your overall program archive. Indeed, you may produce several SAMs in order to depict both different physical areas within a complex infrastructure, and architectural views at different times, for example, before and after integration of the proposed application.

It is important to start off with an accurate depiction of the current systems architecture within your organization. If you do not already have such a document, then BANTAM provides the opportunity to take control of this right away and ensure that someone is tasked with producing an accurate view of the current situation. In the majority of cases, this will be an ongoing task as new developments are introduced and these documents must therefore be kept up to date accordingly. If you do already have a well documented systems architecture, then it will be necessary to identify where the proposed new application will have an impact, and separate this out onto a BANTAM SAM for the purposes of the project at hand. This should be a relatively straightforward task in most instances, although it may turn out to be an iterative process, while the analyst undertaking the task checks the currency of available information against the architectural reality. BANTAM allows for this with its distinction between working and final documents.

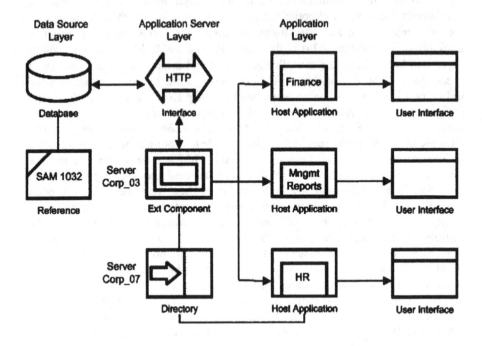

Figure 4.2 Systems Architecture Map graphic

In Figure 4.2 above, a very simple graphic shows the logical relationship between back end data and the application clients, together with the organizational directory which may be used in association with password authentication. Note the reference symbol which indicates that additional detail about the database is available within a separate BANTAM map. Within an actual (as opposed to logical) diagram, applications will be identified by name and more detail will be supplied regarding the network infrastructure. However, Figure 4.2 shows the general concept of using SAMs to describe the underlying architecture, or parts of the underlying architecture. Within a large organization where the systems architecture is likely to be complex, it may be useful to break down the overall picture into logical business area groupings, such as sales, finance, marketing and so on, or perhaps operationally, such as business critical, support, security and so forth. Indeed, you may like to produce various sets of documents in order to be able to describe the architecture accurately from any of these perspectives.

With regard to user authentication and biometric/token technology, you will probably have a high level architectural map in order to describe the underlying principles of operation, complemented by one or more detailed maps to show specific applications and how they sit within the bigger picture. The Map Explanatory Notes (MEN) documents accompanying each of these maps will provide adequate explanation and detail around preferred protocols, available network bandwidth, router components and other relevant details which the systems integrator should be aware of when designing and implementing the system for use within your particular organization and its specific systems architecture.

It will be obvious to the reader, that the use of these SAM diagrams will be of great benefit when discussing the proposed application and how it will sit within your organizational systems infrastructure. Similarly, it will also be obvious that true, realized performance can only be discussed with reference to this architectural picture and the way the proposed application will be integrated into it. It seems almost incredible, but sadly true, that many such systems have in the past been purchased and implemented (often by third parties) without a word about architectural standards or network infrastructure, let alone the use of proper documentation in this respect. Is it any wonder that they often failed to live up to expectations with regards to operational performance? Fortunately, the BANTAM user has the necessary tools to ensure that this is not the case within your organization. A proper architectural plan is an essential component for an efficient IT department. Proper documentation of the same is equally essential from the broader perspective, and especially relevant from the individual application perspective, as will be the case when considering a biometric or token technology based application. The BANTAM SAM should therefore be common currency within your organization in this respect.

4.4 The Logical Scenario Map

The Logical Scenario Map (LSM) represents a particularly interesting element within the BANTAM methodology. It is used to describe a particular operational scenario in sufficient detail to enable the development of the appropriate systems module in order to deliver the requirement. It should be noted that in many cases, the development of the LSM will be a nontrivial exercise which should be undertaken by a suitably experienced team member. Try this simple test. Go to a specific department within your organization, say, finance for example. Choose a function which is undertaken periodically, and ask six different members of the finance team to explain to you in concise terms how the execution of this function works. It is likely, even if they are directly involved in the chosen operation, that you will receive varying descriptions of the function in question. This is simply a product of human nature, people see things slightly differently according to their precise role or the importance that they place on a particular functional operation. When asked unexpectedly to describe it, they naturally describe it from their own perspective and are likely to place emphasis on different parts of the process accordingly. The analyst producing LSMs for your proposed application must get beneath this, to the fine detail of what is really happening within a given operational scenario, and document it accordingly.

Your particular organization may be very good at documenting operational processes and you may already have a certain amount of relative information in this respect. If not, BANTAM provides you with the opportunity to catch up and start following established best practice. Every important operational function within your organization should be properly documented, with the documentation regularly maintained as to its currency. This may sound like a thankless (and costly) task but, in fact, it will pay dividends in areas of training and process re-engineering where appropriate. Furthermore, if you are working towards internationally acknowledged standards of program management, then you should be doing this anyway. For your proposed new biometric and token technology application, it will be particularly important to document the relevant operational scenarios.

A good way of approaching this in many cases will be to document the existing process within a particular operational scenario, and then produce another LSM to indicate precisely how the proposed application will work, thus highlighting the difference within the two documents. This will be especially useful if training or retraining is required for the operational personnel and day-to-day users of the system. It will also be extremely useful for discussions with potential suppliers, in order to show clearly how you anticipate the new operation as a result of the application technology being deployed.

The LSMs may be produced at varying levels of detail according to the complexity of the operation. They will almost certainly be iterative in nature, as the analyst producing the documents tests them against reality. Furthermore, in relatively complex situations, there may be a hierarchy of LSMs, with a relatively high level document referencing several lower level detailed documents as required in order to cover a particular scenario completely. Using the BANTAM symbol notation, they will be easily and quickly produced, enabling the analyst or consultant to progress quickly through iterations to the final document set. When signing off the final documents, the Document Management Group will probably wish to bring in experienced managers from the appropriate operational areas in order to ensure that understanding is both complete and commonly held by all concerned. In this respect, the project is already providing benefit to the operational areas by undertaking this task and giving managers the opportunity to contribute to the process, as well as testing their own understanding around such processes. This will often generate renewed interest in the project overall, and represents another example of how the simplicity and intuitiveness of BANTAM can act as a valuable internal communication tool. Having developed these LSMs, they may of course be reused in a variety of situations, from training and support issues within your own project, to completely separate projects, perhaps run in different departments but affecting the same operational areas. Another example of the value of proper documentation in this context.

In producing the LSMs, the analyst will often start by capturing a single individual's view of a given process and documenting this accordingly. Using this original document in further interviews both with individuals and groups where appropriate, the analyst can quickly refine and elaborate on this original view in order to capture the current process accurately. In doing so, they will also have the opportunity to capture valuable ideas and potential requirements, straight from the people who are closest to the operational scenario in question. Pertinent points in this respect may be noted on the accompanying Map Explanatory Notes, or captured separately if required, in order that they be fed back into the overall program management.

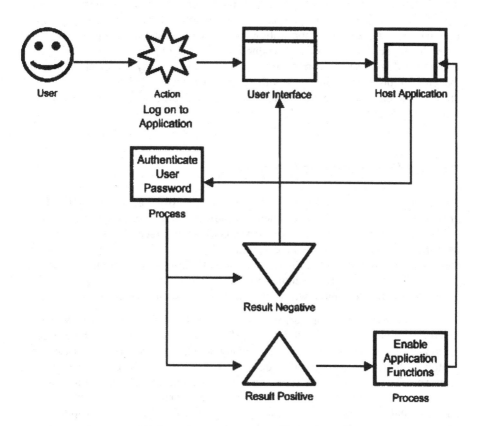

Figure 4.3 Logical Scenario Map graphic

Figure 4.3 above depicts a very simple scenario, showing that the user in this case is currently required to log on to a particular application using a password. An authentication process then takes place, resulting in the user either being allowed access to the application, or receiving a message via the user interface to indicate otherwise. At this level, it is not necessary to show the detail of how the authentication takes place from a systems perspective, as we are simply interested in understanding the current practice for this simple operation. Typically, an LSM will show more detail around the functionality within a given application scenario, together with any outputs as appropriate. In instances where several users are involved, it will also name the users by role, appended to the user symbol. It may also contain several process symbols following on from the core scenario, in order to show the relevance of this operation to other operations further down the chain. These may be broken out into additional LSM documents if they have direct relevance to the proposed new application. Figure 4.3 is deliberately simplistic, in order to illustrate the concept of starting with what may seem obvious, in order to confirm the current understanding and be able to move forward into designing and documenting the new processes, which will in turn constitute the operation of the new application and will therefore have a direct influence on the functional design of this application. Within your master project plan, adequate time should be allocated to the production of the LSM final documents. It is imperative that these are correct in their detail and that they accurately represent the broader program aspirations with regard to the proposed new processes, as these documents and the conclusions drawn from them will influence all that follows.

4.5 The Functional Scenario Map

The Functional Scenario Map (FSM) will often mirror an associated LSM by providing the necessary functional detail to support the process in question. There will be several FSMs within a typical project, and together they will effectively describe the application in technical terms. In a particularly complex situation, there may be a hierarchy of FSMs, systematically drilling down from a high level view to the level of fine detail necessary to develop the application. The BANTAM reference symbol will be the link between such a document set.

You may reasonably ask why we need an FSM as well as an LSM? The two documents will be produced from a different perspective and often by different personnel working on the project. As already noted, the LSMs are concerned with capturing both the current and future processes around a given operational scenario. They are not concerned with the technicalities of supporting such a scenario, but rather concentrate on the logical flow from a users perspective. The business analyst or consultant producing the LSMs will have a good grasp of these processes from an operational perspective, but will not necessarily be a technically biased application developer. The FSMs will be more technical in nature and will show the path forward for the development or integration of the proposed application. These documents will need to be produced by someone who understands the technicalities involved and is capable of developing solutions which will support the original aspiration. This will be undertaken with full reference to the SAMs as well as the information provided by potential device or software suppliers. Using a series of FSMs in order to describe the overall situation in manageable chunks is an efficient methodology for application development purposes. If these individual sections match closely to operational processes, as they will when BANTAM is used, then so much the better, as this will promote clarity and understanding across the board and ensure that the delivered application closely matches the original requirement. Making full use of the SAMs and LSMs when developing the FSMs will also ensure that no time is wasted following incorrect assumptions and that the team members producing the FSM documents (who may be a mixture of internal and external personnel) do so with full knowledge not only of the requirements, but also the technical infrastructure within which they are to be delivered. Now we are really seeing some value from BANTAM!

An FSM may sometimes be quite complex in nature, especially when dealing with interfaces between components. For this reason, even a single FSM may be abstracted out to additional maps in order to describe a particular interface or protocol. Similarly, within a BANTAM FSM, reference may be made to an external document or technical paper in order to elaborate on areas of connectivity where this is considered pertinent to the application overall. Indeed, part of the power of BANTAM is the ability to describe every element of an application design, down to the very lowest level of detail where necessary. A program or application designed using the BANTAM methodology should thus be documented in such a way that the entire application could be reconstructed if necessary by reference to the archived documentation. This emphasizes the concept of repeatability and reuse of system components, a concept embraced by BANTAM. If you are managing a series of projects within a large organization or government department, a considerable amount of time may be saved by learning from previously successful implementations of different functional elements. The FSM provides the opportunity to do exactly that. Furthermore, such maps may be incrementally refined from project to project, strengthening the knowledge gained from practical experience and leading to best practice in implementation. If you are a systems integrator or supplier, you will readily appreciate the value of such an approach, both commercially, and from the customer service perspective. Documenting an application in this manner will also pay dividends in regard to subsequent maintenance which, after all, may be undertaken by a third party who was not involved with the original application design.

In many ways, the FSM may be considered to be at the heart of BANTAM, providing the nuts and bolts detailed view of exactly how a given application works and further providing a mechanism for actively designing elements of an application which is treading new ground in areas previously untouched by this technology. Bringing clarity to this area is of vital importance if we are to realize a successful project implementation.

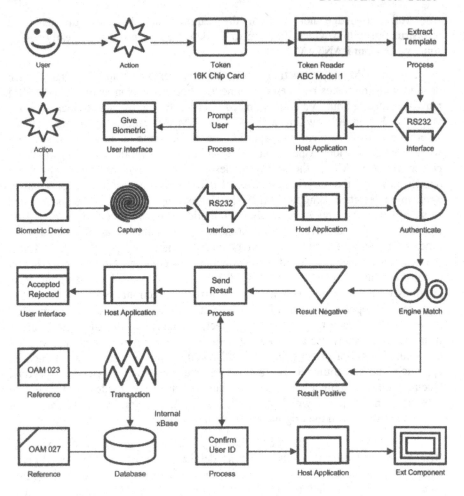

Figure 4.4 Functional Scenario Map graphic

The FSM depicted in Figure 4.4 shows the simple process of authenticating the user and, if the claimed identity is verified, passing a message to an external component, which in most cases will be another application. Even at this relatively high level, one can see how the system is working, or required to work, and can discuss the finer details accordingly. In this example, the user presents his or her chip card to the token reader, which extracts the template information and passes it to the host application via a known interface. The host application then prompts the user to provide their biometric sample, which is subject to authentication against the reference template. A decision is made via the matching engine and the result communicated back to the user interface via the host application. Upon a positive result, further communication takes place with an external component.

Beneath this high level FSM there will be additional detail. For example, if I were developing the host application, I would now know that I will receive communication from both the token reader and biometric device via an RS232 interface. This tells me that I need to constantly monitor the RS232 communication port buffers on the host PC, however, as yet, I do not know the precise data format and so I cannot parse the data stream in a meaningful way. The next level of detail within a lower level FSM will no doubt provide this information for me. In Figure 4.4 you will notice that both the transaction symbol and the database symbol each point to a reference symbol, which tells us that another BANTAM map provides a lower level of detail about these components. This hierarchical methodology allows us to develop the required functionality within our application systematically in an intuitive and easy to follow manner, obtaining confirmation at each stage, and ensuring that the right functionality is being developed in the most appropriate manner. In many projects, there may only be two or three FSMs; it doesn't really matter how many documents you use, the main point is that you are able to elaborate the detail as much or as little as required in order to reach the desired conclusion. Importantly, it also provides for a documented record of how you arrived at this conclusion. From a project management perspective, it allows developers to work in a logical fashion, addressing manageable chunks which may be allocated sensible time windows accordingly. It also highlights any problem areas in good time and draws attention to potential weaknesses in either infrastructure or systems design.

Professional developers will respect and welcome this approach. Project managers will be empowered with quality information at all times, and the organization will get the application it was expecting. These are pretty good reasons for introducing a methodology such as the BANTAM FSMs. The benefits don't stop there however. By the time your application is ready for implementation, you will have a wealth of detailed and valuable information concerning every aspect of the application. When it comes to designing a training package, or putting together a tender for subsequent maintenance of the system, there will be no question about exactly how the system is operating. Similarly, if there are questions around the potential impact of your application upon other systems, all the necessary detail is there for reference. The BANTAM FSMs will represent a most valuable element within your overall program management and will repay many times over the effort expended in creating them.

4.6 The Object Association Map

The Object Association Map (OAM) has a very particular role to play within the overall BANTAM methodology. At first sight, a BANTAM OAM may look familiar to those conversant with entity relationship modeling for database applications, and certainly the OAM can function in exactly this role. However, the OAM in BANTAM is designed to be more flexible than this, covering anything which is an entity or object, whether based in software or hardware. For example, an object may well be a data entity such as 'user' which has attributes such as 'name' or 'staff number'. In this context, 'user' may translate into a database table with 'name' and 'staff number' becoming fields or columns within that table, which may have relationships with fields in other tables. On the other hand, an object may be a chip card whose relationship or association is with a reader, interface and host application, between which data is passed according to a particular protocol. This data, in turn, may have relationships with other data entities, producing a slightly more complex picture, but nevertheless, one which is important to understand. The primary purpose therefore of the BANTAM OAM is to illustrate associations between objects, whatever those objects may be, and however is necessary, in order to describe a particular scenario adequately.

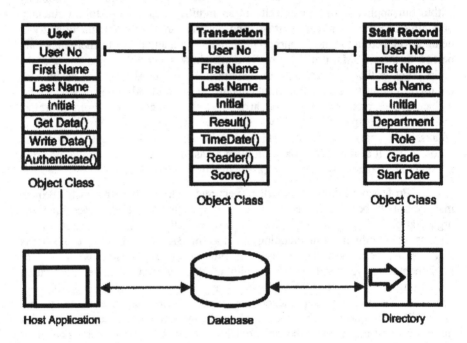

Figure 4.5 Object Association Map graphic

In Figure 4.5, the OAM is used within a software context, to illustrate the relationships between a user record stored within the host application, a transaction record stored within a central database and a staff record stored within the organizational directory. In this particular example, there is a certain amount of data duplication which may have been deliberately configured in the interests of performance, and the OAM shows this clearly. Also in this example, the physical objects with which the data is associated are also shown. We may have shown references to other BANTAM maps if we wished to further align this view with, for example, an architectural map.

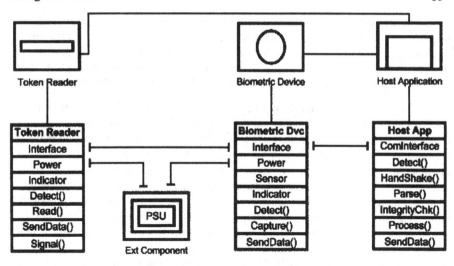

Figure 4.6 Object Association Map graphic 2

In Figure 4.6 above, the OAM is used in a slightly different manner in order to illustrate a common interface between hardware components and the host application. It shows the primary attributes and functions of the components in this context and additionally indicates a common power source for the reader devices. This is a simplistic example in order to illustrate the concept. In reality, we may have chosen to elaborate this further with precise details of the interface protocol and data element formats. We may also have appended additional detail around power requirements, current draw and so forth for the physical reader components.

The OAM may therefore be used to depict any scenario wherein it is important to understand the relationships or associations between objects. Mostly, this will be from a software transactional perspective where the objects will be data elements. However, it may occasionally be useful to extend this to hardware components or even mix elements of both on the same map. The BANTAM symbol library allows you to do this easily, and the hierarchical nature of the BANTAM documents allows you to elaborate down to whatever level of detail is required, via the reference symbol and a logical progression of related documents. In this respect, the OAM maps will often be referenced from different types of BANTAM maps such as the SAM or FSM maps.

4.7 The Miscellaneous Definition Map

The BANTAM document set has been designed to cater for most eventualities found within a typical biometric or token technology project, as well as allowing for a logical progression from aspiration to implementation. However, in designing the BANTAM document set, a conscious effort was made to keep the number of documents constrained in the interests of simplicity. One must acknowledge also that, however many documents you include in the methodology, there will always be some scenario or other that doesn't quite fit the model. For this reason, BANTAM includes a Miscellaneous Definition Map (MDM).

The BANTAM MDM may be used in any situation where the other maps don't quite cover the eventuality, and yet it is considered important to capture either a process or area of systems functionality. For example, it may be that we wish to describe a part of another system which isn't a direct part of our application, and over which we have no control, but that nevertheless has some relevance to our situation. Or perhaps, as a supplier, we wish to document another system which we believe has some synergy with the application being proposed, and from which there may be learning points. Alternatively, we may not be capturing a systems related scenario at all, but perhaps an external process which occurs before or after the functionality of our application and may therefore have some relevance. Indeed, the MDM is an extremely versatile document which may be pressed into service for a variety of purposes. However, it still uses the BANTAM symbol set and all the attendant conventions, such as the reference symbol to point to other documents and so on. It also has its own Map Explanatory Notes to elaborate upon the graphical representation within the MDM, and these are completed in the same manner as with any of the regular BANTAM documents.

Perhaps even more importantly, the MDM is considered an integral part of the overall BANTAM documentation and, when used, forms an essential part of the program archive. If an MDM was considered important enough to generate in the first place, then it is equally important to capture this within the program archive, in order that it may be referred to after the event where necessary. The combination of the primary BANTAM map documents plus the MDM, makes for a truly comprehensive approach to application definition, design and implementation, within a structured and logical framework.

4.8 The Map Explanatory Notes

As previously mentioned, every single BANTAM map document is accompanied by a Map Explanatory Notes (MEN) document. Whenever you see a standard BANTAM map, turn the page and you will find an MEN to elaborate in plain language upon the associated map. This is important, as there may be particular issues or circumstances which need to be explained in relation to the process or function being depicted. It also provides the map author with the opportunity to specify individual components where appropriate and offer any general observations considered pertinent to the situation at hand. In many instances, it may be thought of as the qualification of the graphical representation within the main BANTAM map. Indeed, the BANTAM map author will often find it useful to complete the two documents in parallel, ensuring that every important element is covered.

The MEN document must be completed for every primary BANTAM map and stored together for easy reference. The MEN template includes a provision to reference the associated primary map. The two documents are subsequently read together when using the BANTAM maps either for general discussion or for specific development purposes. This methodology ensures that the majority of individuals involved with a given program or project, can easily understand what is being described within the BANTAM maps, regardless of their individual technical understanding. This is a cornerstone of the BANTAM approach and must be adhered to if the full benefits of BANTAM are to be realized.

Rather than taking a completely unstructured approach, the MEN document is divided into three sections, in order to provide clarity and consistency across the program and between different working groups, be they internal or external.

The Map Overview section is where the map author may provide a general overview of the process or function being described within the master document. In this section, the author may explain why a particular approach is being taken, either in process or architectural terms. Within this section, we may also choose to elaborate upon any external component symbols used within the main map, if they can be simply described in plain language. If the external component is considered complex enough to warrant a break out map of its own, then that map will have its own MEN document in order to describe the detail. Also, within the Map Overview section, we may identify different departments within the organization who have a stake in the process or function being depicted. This sort of information can be extremely valuable after the event when subsequent project teams are required to interface with or otherwise enhance the application. In short, any pertinent information which directly supports or qualifies the primary BANTAM map document may be included within the Map Overview section of the MEN.

The Component/Device Details section provides an area in which the author may list details of specific hardware or software components which are being considered for the application and are therefore referenced within the primary BANTAM maps. For example, we may be anticipating the use of a specific biometric capture device, or a particular type of chip card within our application. Alternatively, there may be existing devices which we wish to incorporate within the application, and which therefore need to be identified accordingly. Also in this section, we may list the fundamental software building blocks, such as the type of database being deployed, the network operating system, the client operating system, the type of web server being deployed and so on. Once again, this may be a mixture of new and existing technology. For both hardware and software items included in this section, the author may usefully decide to include the suppliers contact details (as understood at the time) in order that subsequent users of the documentation may know who to contact for additional information or support.

The Relative Information section is the map author's scratch pad, where he or she may capture any information considered pertinent to the overall picture, in a relatively unstructured manner. This will be especially useful in situations where the author has a strong point of view about a particular item which he or she wishes to log with the primary document. For example, it may be that a device anticipated for use within the application has not been formally tested by the organization and therefore its inclusion is based purely upon published specifications. Or it may be that the design approach taken was acknowledged as suboptimal at the time, but taken for other reasons of expediency. This type of information is extremely valuable further down the line, and we must provide a mechanism for capturing it. Similarly, the map author may wish to record an instance of above average technical knowledge, support or general enthusiasm, experienced from an external entity. The Relative Information section may, in many instances, contain very important information relating to the project, which, without a methodology such as BANTAM, would otherwise have been lost. Bear in mind also that these days, more than ever, company personnel are mobile, and the team who designed the original application may not all be available when it is time to revise or enhance the application, making you somewhat reliant upon the quality and detail of your original documentation.

In conclusion, the MEN document is an important part of the BANTAM methodology. It provides the plain language explanation and detail necessary in order to interpret the graphical depiction within the primary BANTAM map correctly. In addition, it aids communication and discussion within and between project teams, as team members may read the notes in advance of meetings and can therefore be more concise as to their observations and comments accordingly.

5. Associated documentation

The primary BANTAM map documents provide a comprehensive mechanism for defining and describing the various operational processes and systems functionality, which together constitute a biometric, token technology or related application. In previous chapters we have described these documents and their purpose, together with an overview of the BANTAM symbol library and how it may be used to create the various BANTAM maps. The methodology described up to now, already provides considerable value in relation to a typical program or individual project. Useful as this may be, we must also think about the bigger picture and how BANTAM may be integrated within your overall program. This may include elements of traditional project management, procurement, training and other areas, depending on the nature of your organization and the scope of the program being undertaken. BANTAM takes these issues into consideration and provides additional support for these key areas if and as required. BANTAM is therefore pertinent throughout all stages of your program, from the original idea, through application development, procurement and on to implementation. In fact, it doesn't stop there, as you will be able to make use of the BANTAM documentation for both user and administration training purposes, ongoing systems maintenance and support, and much more. Indeed, BANTAM will prove to be a valuable partner for all your biometric or token technology and related projects.

5.1 Requests for information or proposals

An important element within any program is researching available technology and requesting information from potential suppliers. This needs to be handled methodically and dispassionately if you are to identify and select the right technology for your project. It is very easy to get carried away with impressive looking sales material and fancy presentations, and while there is a lot of great technology out there, there is only one combination which is just right for the application you have in mind. Getting to this right combination takes a little effort, but it is effort which will be repaid a hundredfold when the application finally goes live and is being used in earnest.

We have already discussed the primary BANTAM maps and how they can help you to identify current and proposed processes, define systems functionality, describe the architectural requirements and more. When you have gone through the various phases involved in the production of these maps, what you will have will be a comprehensive picture of your current systems and processes, an equally comprehensive description of your requirements for the new application, and a detailed plan of how you intend to implement such an application. You may be able to undertake all of these tasks in-house, or you may have solicited the help of external consultants. In any event, there will be points along the way where you will need more information about the available technology and how it might fit into your proposed application. Fortunately, the fact that you have documented your requirements and current situation to an advanced degree using BANTAM will be of enormous help to the technology suppliers you contact, as they will be able to see immediately what you are trying to achieve and, therefore, be able to advise exactly how their particular products or services might fit. Furthermore, you should find that the majority of established vendors in this technology area are well aware of BANTAM and will be able to respond using the same methodology.

The document which facilitates this is called the BANTAM Request for Information (RFI). It is a simple three-page document which, in turn, will reference the BANTAM maps relevant to the particular piece of technology you are seeking. It is deliberately structured in a simple manner in order to capture the important information and bring consistency to the process, making it easier for you to evaluate responses in a fair and equal manner. The first section of the document gets right down to business and explains exactly who you are, who the respondent should contact, when a response is required, and gives the appropriate reference.

Respond to		Date of issue	19th Jan 2001
Organisation	Department of Defence	Response req. by	28th Feb 2001
Name	Bert Higgins	Reference	DOD 0001
Title	Project Manager		
Address	3101 Victoria Avenue		
Town	Any Town		
County	Any County	Telephone	0198 652876
Post Code	AA1 001	Fax	0198 652877
Country	Any Country	E-mail	bert@dod.mil

Figure 5.1 The RFI header section

Figure 5.1 shows this first section of the document and how it clearly gathers all the pertinent information together for the benefit of the respondent. This is important. We do not want the respondent to have to wade through a wad of pages trying to discover who he should be speaking with, or where to send his response. Especially if there is a possibility that he could become confused and send it to the wrong department or individual, or simply address the package incorrectly. Let's make things easy for all concerned by stating these important points clearly, right at the top of the document where they can be easily seen. The next part of the document categorizes the broad technology area (remember, the organization concerned may be active in several areas of technology) and provides a summary of the BANTAM maps which are included with this RFI. The recipient can thus easily check that all relevant information is received before constructing a suitable response. When a given supplier organization starts to see more BANTAM RFIs and becomes familiar with the process, it will also become increasingly efficient in responding to them, saving time and promoting a higher level of professionalism throughout.

Product / Service Category	Biometric Authentication System

Bantam Maps Included in RFI	
Document Title	Document Reference
Project concept	ALM 003
Operational logic	LSM 003
Existing infrastructure	SAM 003

Figure 5.2 The RFI BANTAM map summary

In Figure 5.2 above, we can see how the RFI document clearly lists all the attached BANTAM maps, provided in order to describe the requirement clearly and help the respondent to understand the infrastructure into which their technology might be integrated. The broader product or service category is also listed, enabling the recipient to field the RFI quickly to the most appropriate department or individual within the organization who may best respond to this inquiry.

The next sections of the BANTAM RFI provide free-form text areas for the following purposes. The Basis Overview provides a brief summary of the RFI and the basis under which it is issued, together with a high level overview of the proposed application, as far as is necessary for the respondent to understand fully what information is being requested and why. The Perceived Requirement section spells out exactly what is required in plain language terms, leaving no room for ambiguity or misunderstanding. The Special Conditions area outlines any special environmental or other conditions which may impact the application overall and the technology or product being inquired about in particular. It may also outline any expected performance or durability criteria, expectations of support services or anything else which your organization believes a potential supplier should understand about this project. The About Our Organization box provides, as the name suggests, an area where you may describe your organization and its primary activities. This helps the recipient to get a feel for the application in broader terms and how you might wish to implement it. The next section is entitled Format of Response Required, and clearly sets out how you wish to receive a response. For example, you may ask explicitly that certain BANTAM maps be produced in order to show exactly how the supplier would recommend the integration of their products into your application. Or perhaps how many copies of the response are required, in what media format and so on. Standardizing these details will make it much easier for you to evaluate responses when they are subsequently received. Lastly, the Procedure for Questions section explains clearly how a potential supplier should field any questions they have concerning this RFI. This is an important area, especially for large organizations where you may not wish to receive random questions, or have the wrong individuals approached about such matters. At the end of this section of the document, we have a signatory area which clearly indicates who has authorized this particular document.

Signatory to this RFI

Name	Arnold Trubshaw
Department	DOD / BBM 4
Title	Commander
Date Signed	19th Jan 2001

Signature

Figure 5.3 The RFI authorizing signatory

This, together with the information contained in the header section, leaves the recipient in no doubt as to the origins of this particular RFI. We mentioned that the issuing signatory area comes at the end of this *section* of the RFI. In fact, there is more to the RFI, as it also contains a Response Summary in order to capture pertinent information about the respondent organization consistently, making it easier for you to compare like with like and understand exactly who you are dealing with. Professional supplier organizations will welcome this level of detail and clarity within the RFI documentation.

Organisation	TurboTech Inc.	Core Business	Biometrics
Principal Contact	Miles Smythe	Years in Business	5
Title	Senior Consultant	Annual Turnover	$7m
Address	Unit 13 Tech Park	Registered No	32875639
Town	Any Town		
County	Any County	Telephone	0736 372987
Post Code	LKJ 005	Fax	0736 372988
Country	Any Country	E-mail	miles@turbotech.com

Figure 5.4 The respondent details summary

The respondent details requested in the dialog depicted in figure 5.4 are perfectly straightforward and show you precisely who you should be dealing with within the supplier organization, in addition to giving you an indication of how well established this particular organization is within this market sector. The remainder of the Response Summary provides a section where the respondent may note any other pertinent information which they would like you to consider with respect to this RFI, a summary of the BANTAM maps included with the response, and a summary of other attachments, such as product literature, white papers and so on. Lastly, the response summary includes an authorized signatory to the response, who should be a suitably qualified individual within the supplier organization. You will appreciate that this standard and structured approach to issuing RFIs will provide many benefits while you are gathering information as to relevant technology or services for use within your application. However, there will come a point, further into your project, when you will wish to escalate this into firm proposals for the supply of such technology or services. BANTAM has this covered also.

The Request for Proposal (RFP) is an important document within the BANTAM portfolio. The content of this document and the responses received as a result, could have a dramatic affect on your final application. Having undertaken all the good work in specifying operational processes and designing high level systems functionality, we now have to put this to the test by ascertaining exactly how the application will be implemented, and precisely which technology will be used. Time spent on producing the RFP and managing responses accordingly, will be time well spent in the context of the overall program. The BANTAM RFP makes this easy for you, with the provision of a clear and concise standard document which covers all the fundamentals and allows you to elaborate where appropriate.

Respond to		Date of issue	19th Jan 2001
Organisation	Department of Environment	Response req. by	28th Feb 2001
Name	Molly Clackett	Reference	DOE 0003
Title	Project Manager		
Address	1016 Albert Avenue		
Town	Any Town		
County	Any County	Telephone	0123 765378
Post Code	AA1 001	Fax	0123 765379
Country	Any Country	E-mail	molly@doe.gov

Project Name	Remote Authentication Project

Figure 5.5 The RFP header information

Figure 5.5 above shows that the RFP header information is virtually the same as that for the RFI document, except that the project for which a proposal is being sought is now explicitly named. The remainder of the RFP document however is quite different from the RFI, as it has a distinctly different purpose. A BANTAM maps summary is provided, followed by a Project Overview which offers a plain language description of the project and its broad objectives. As much space as necessary may be taken for this overview, as it is vitally important that the respondent fully understands your aspirations in this context. At the end of this, and following sections, the RFP has a single line footer which provides

BANTAM map references for what has just been discussed, allowing the respondent to follow logically through the document and check their understanding at each stage.

Map Refs	SAM 005	ALM 031	FSM 017		

Figure 5.6 The RFP Map Ref. footer for different document sections

The next section is the Detailed Requirements section, where, as the title suggests, you may spell out the detail of exactly what you would like the respondent to produce a proposal for. This may range from a small area of functionality, to the complete program, depending upon your situation and how the application has been designed. Once again, this is a plain language description of exactly what you are looking for. The technical detail will be provided in the attached BANTAM maps and referenced both in the footer to this section and the overall map summary. There is therefore no need to get too technical in this description, it is much better to provide a succinct and highly readable description which leaves the respondent in no doubt over what they should be proposing and why.

Special Operational/Architectural Conditions is the title of the next section of the document, and it represents a very important part of the RFP. Here is your opportunity to draw attention to the particular operational conditions at the site in question. This may include numbers of users and expected transaction clustering within a 24 hour period. It may include unusual environmental conditions, such as excessive temperature swings, humidity, or exposure to bright sunlight. It may include difficult environmental conditions such as high levels of dust or dirt, excessive noise and other conditions. It may include administration expectations and available resources, and so on. From an architectural perspective, we may draw attention to any network specific situations, such as capacity and performance issues within a given time cycle, or any unusual security architectures currently in place which need to be taken into consideration. The attached BANTAM maps will cover the fine detail of this, but we need to make quite clear to the respondent any unusual conditions which may affect the feasibility of their proposal. This is not only fair to them, but makes life easier for you further down the line.

In many instances, the respondent will wish to make a site visit in order to appreciate the situation for themselves and check their understanding of the operational conditions and your broader aspirations. This is quite reasonable and should be encouraged, as you will want potential suppliers to have a good understanding of your situation and how you envisage the application will work. However, it is important that the question of site visits is managed properly and that all respondents are treated equally and have access to the same information. You therefore need to think about this in advance and make sure that everything is in place to accommodate such requirements before you formally issue the RFP.

BANTAM helps you achieve consistency in this respect by providing a site visits section within the standard RFP document.

Site Visits

Available Yes / No	Yes	Preferred Days	Mon - Wed
Contact Name	Rodney Williams	Preferred Time	10.00
Department	Facilities	Notice Req.	5 Days
Title	Manager	No. Of Visitors	3 Maximum
Telephone	0187 363578	On Site Parking	Yes
E-mail	rod@doe.gov		

Related On Site Requirements

This is a hard hat environment and appropriate safety clothing must be worn at all times.
Security passes must be worn at all times and will be issued at reception.
The health and safety emergency procedures document must be read by all visitors to the site. Copies are available at reception.
Visitors vehicles must display a visitors badge if parked in the main car parks.

Figure 5.7 The site visits section of the RFP document

This simple dialog provides clear information to the respondent on what they need to do to organize a site visit, and any related details which need to be taken into consideration. A single point of contact is provided, ensuring that all respondents will be advised in the same manner and treated equally when on site.

As with the RFI document, About Our Organization and Format of Response Required sections follow, allowing you the opportunity to share appropriate information with respondents and clearly state how you would like them to respond. If you encourage them to respond using the same BANTAM methodology, then there is a clear advantage, as you will be able to make comparisons between relevant BANTAM maps and quickly understand not only what is being proposed, but exactly how the respondent aims to integrate with your particular infrastructure. Working in this manner can save a considerable amount of time for all concerned and ensure that important details are not overlooked.

The next two sections of the RFP are particularly important for both parties. Firstly, the Procedure for Questions section sets out clearly how an approach should be made if the respondent has additional questions which he considers are not covered within the RFP. It is unlikely that there will be many of these, but, nevertheless, we must provide a clear mechanism for dealing with them. What we don't want, is random contact being made with the wrong personnel within your organization, which may be difficult to track and manage properly. Similarly, we don't want any 'back door' approaches by potential suppliers seeking to gain some competitive advantage via familiarity or intelligence gained in this manner. Indeed, you should have a procedure for dealing with such a situation. Consistency and strictly adhered to procedures represent the fairest policy for both parties. You should thus make clear the procedure for additional questions within the RFP document, and BANTAM provides the mechanism to do just this. Secondly, the How We Shall Evaluate Responses section sets out clearly the internal procedure which shall be followed in order to evaluate responses. This may take the form of a project board meeting in order to match responses against the requirement and derive a shortlist accordingly, or perhaps some other mechanism. In any event, it is good practice to state this clearly, in order that respondents fully understand how decisions will be made. You may also include in this section the next logical steps in the procurement process. For example, you may decide that after a shortlist is generated, potential suppliers will be invited in for a formal presentation of their proposal. Again, it is good to share this information with respondents, in order that everyone knows exactly where they stand. Finally, in this part of the RFP is the authorized signatory dialog where the appropriate individual from your organization signs and dates the RFP.

As with the RFI document, the RFP contains a Response Summary in order to bring some consistency to the way potential suppliers respond to your RFP. This starts with a header to capture the pertinent high level details of the respondent's organization.

About the Respondent

Organisation	Systems Designs Inc	Core Business	Networks
Principal Contact	Eddie Wilson	Years in Business	3
Title	Manager	Annual Turnover	$13m
Address	1239 Indian Drive	Registered No	398765
Town	Any Town		
County	Any County	Telephone	703 256 789
Post Code	123 768	Fax	703 256 790
Country	Any Country	E-mail	ed@sysdesign.com

Principal Company Officers

Name	Title	Telephone
Rick Farnsworth	CEO	703 987 456
Charles Forrera	VP Finance	703 987 457
Gail Williams	VP Marketing	703 987 450

Figure 5.8 RFP Response Summary header

You will see from Figure 5.8 that the response summary header also includes details of the principal company officers. It is as well to know exactly who you are dealing with. This is followed by an Other Pertinent Information section wherein the respondent may add other relevant information concerning their organization, in the context of this RFP. There follows a section, as illustrated below, where the respondent may supply details of similar projects undertaken, together with a reference site which they are happy for your organization to contact in relation to this project.

Similar Projects Undertaken

Client	Description	Date
MHS	Network update	10/02/1999
Traintrack	Signalling system	03/05/2000

Reference Site Details

Organisation	ABP	Contact Name	Roger Dowd
Address	ABP House	Telephone	508 657 382
Town	Any Town	E-mail	roger@abp.com
County	Any County		
Post Code	AA2 98P	Type of Project	Network
Country	Any Country	Date Completed	11/09/2000

Figure 5.9 RFP Response Summary reference section

If the respondent's proposal is of interest, the RFP issuing organization may well wish to contact one or more organizations for whom the respondent has undertaken similar project work. Naturally, this should only be done with full agreement from the respondent, who, in most cases, will be pleased to have the opportunity to show examples of previous work and have former clients describe the application and how it has been of benefit, or even discuss any perceived pitfalls encountered during implementation. This sort of open communication is in everyone's interest in the long term.

The next section contains a summary of the BANTAM maps included with the response, together with a summary of other attachments. This helps the issuing organization to check that all the relevant documents have been received and that they can make a fair appraisal of the response accordingly. Remember, this is a simple two page response summary, which mostly will be used as a checklist and focus for discussion during the RFP evaluation phase. The detail of the response will be captured within the accompanying BANTAM maps and other attachments. In a similar vein, the next section provides a high level summary of costs. The detailed breakdown of these costs will be provided elsewhere, but the summary allows easy comparison of the pertinent points, such as the distinction between equipment and installation costs, and allows the respondent to draw attention to any 'other' significant costs which need to be taken into account.

Estimated Cost Summary

Estimated Equipment Costs	€20,000
Estimated Installation Costs	€50,000
Estimated Other Costs	€10,000
Total Estimated Cost	€80,000
Currency Expression	euro

Detail of Other Costs

Project management and documentation

Figure 5.10 RFP Estimated Cost Summary

The last section within the RFP Response Summary is the authorized signatory dialog where a suitably qualified individual from the respondent's organization can sign off the response. The returned BANTAM RFP, with properly completed Response Summary and an appropriate set of BANTAM maps with which to describe exactly what is being offered and how it might be implemented, will provide everything you need in order to make a sensible evaluation of the proposal in question.

5.2 Response to requests

In the previous section, we looked at RFIs and RFPs from the perspective of the issuing organization or end user and explored the documents themselves. In this section, we shall consider responding to RFIs and RFPs from the perspective of the suppliers or manufacturers.

When you receive an RFI or RFP from an organization using the BANTAM format, it is important to study the attached BANTAM maps carefully and ensure that you understand what is being described and the solution being sought. The first thing will be to check the BANTAM map summary and ensure that you have all the documentation. You might then usefully read the overviews and any information supplied about the organization in question, in order to get a feel for the overall situation. If the supplied BANTAM maps extend down to an architectural or detailed technical level, then it is important that a suitably qualified member of your staff appraises these maps and liaises with the team accordingly. In such a case, do not let sales staff respond to the RFI or RFP without this input (unless, of course, they happen to be technically qualified and experienced). In any event, a response to a significant RFI or RFP should be a team effort. On no account allow secretarial or administration staff to simply mail back a brochure and compliments slip, this is the very worst kind of response to an intelligently conceived RFI or RFP.

Naturally, in order to respond fully and in a like manner, your organization should be familiar with BANTAM and able to produce the full set of BANTAM maps. You will find this capability of enormous benefit, not just in order to be able to respond to BANTAM RFIs and RFPs, but for general use in bringing clarity and consistency to your project proposals, however they originated. But there is another valuable aspect to this from the supplier's perspective. Using BANTAM consistently for all your project proposals and designs will mean that you automatically develop a technical archive, outlining the way you approached various challenges and successfully developed solutions accordingly. This is not only invaluable for training purposes, but promotes the concept of reuse, as you can learn from successful projects and reuse proven technical solutions in subsequent programs. This is a powerful working methodology, with the potential to save considerable time and effort across your portfolio of biometric and related projects.

In responding to BANTAM RFIs and RFPs, ensure also that you use and fully complete the Response Summary sheets provided. This makes it much easier for the organization in question to evaluate your response properly and also demonstrates your familiarity with BANTAM. It is a good idea to comment on any user supplied BANTAM map and, where appropriate, to respond with an equivalent map which details your own response and suggestions accordingly. Certainly, you may wish to support your response with relevant product brochures. You may even choose to reference these brochures from within BANTAM maps, if it helps in providing technical detail to support a proposed design element. It is important to be able to demonstrate that you have fully understood the requirement and have carefully considered your proposed solution. The use of BANTAM facilitates this and enables you to provide as much detail as is necessary in order to describe your proposal. Furthermore, assuming you are invited to meet with the end user, you will find that the maps you prepared for your response will act as valuable discussion documents. You may like to have enlarged versions printed for just this purpose, or perhaps to incorporate the BANTAM maps in an electronic slide show in order to present your proposal in sufficient detail (in fact, a template is provided on the accompanying CD-ROM for just this purpose).

In conclusion, as a supplier or systems integrator, you are encouraged to embrace the BANTAM methodology and make full use of its potential, not only for responding professionally to BANTAM RFIs and RFPs, but for all the additional benefits it can bring to the way you work within your organization. You will discover many such benefits when you standardize the use of BANTAM for all your solution development and documentation requirements.

5.3 Program management

This section will explore how the use of BANTAM can greatly enhance overall program or project management within your organization but first, we may like to define exactly what we mean by these terms.

For many organizations, a program represents a broad initiative aimed at improving a particular organizational process, or maybe even the fundamental operation and configuration of the organization itself. Within this broader program, there may be a number of individual projects, each of which will deliver a defined operational benefit. Both programs and projects tend to be scaleable from minor departmental operations to significant organizational change. There are of course many software applications and techniques available to program and project managers, who will no doubt have their preferred methodology for tracking projects. Such techniques often focus on project milestones, deliverables and resource management, with rather less focus on the quality of delivered solutions or the alignment with original objectives. There is thus a need in many cases for supplemental processes to support the higher level milestones with quality analysis and design. There is also a need for proper documentation in the form of a program or project archive, containing all the relevant process and system design documents used in the various stages of the project. BANTAM provides this capability.

A good place to start integrating BANTAM into your overall program management, is to ensure that the high level application logic and systems architecture maps appear as milestones early on in your project plan. The various Logical Scenario and Functional Scenario maps will appear across the project timeline as different application areas are addressed. Similarly, other BANTAM map documents will appear as necessary throughout the project life cycle. In each case, we should not only identify the need for a specific map, but designate a time frame for its completion and monitor progress towards this accordingly. One additional benefit of working this way, is that you can reference the BANTAM maps by name and number, making it easy for the project team to find specific and detailed information appertaining to a particular phase or activity within the project. You may also wish to incorporate the management of RFIs and RFPs into the project life cycle, referencing the appropriate BANTAM documents as appropriate and tracking their issuance and responses received. In this respect, the BANTAM methodology fits comfortably with typical program management techniques, providing an additional dimension in aligning the detailed BANTAM documentation with phases, milestones and activities within the overall project plan. This provides for much stronger tracking capabilities, especially when a project is being subsequently evaluated, perhaps in an effort to capture best practice. It also provides the project manager with a much more realistic view of what is actually being discussed and worked on at a given time.

In addition to the above, BANTAM provides other enhancements to program management, especially with regard to documentation and project archives. The BANTAM document set has been designed to be easily managed and grouped for specific projects, allowing for logical document archiving at both project and program level. If your organization is working to internationally accepted quality standards, then the creation of detailed project archives will form an essential part of your overall working practices. In addition, there may be substantial opportunities for reuse and associated cost savings, by adopting a working routine which checks for similarities across projects, whether at system component or process levels. The clarity and detail provided by BANTAM makes this easy to achieve with a little thought. You may also wish to integrate elements of BANTAM into your resource management methodology. For example, if you are using a time recording system to track resources against projects and project activities, then creation of the primary BANTAM maps and preparation of RFPs may be useful items to include in your task lists for given individuals.

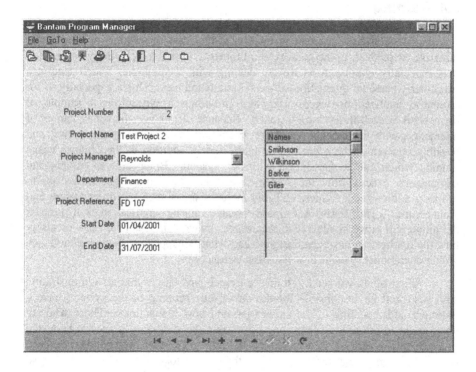

Figure 5.11 The BANTAM Program Manager

If you are undertaking several projects as part of a larger initiative, or perhaps responsible for multiple projects across different departments, then you may wish to use appropriate software tools to help you manage such a situation. This will be especially pertinent with regard to document management and resourcing issues. You may already use tools to help you in this respect, but, if you do need some additional coordination, the BANTAM Program Manager (a demonstration version of which is on the accompanying CD) may prove to be useful, providing an intuitive and simple way to track project related documents as well as a helpful database for personnel and external suppliers.

There are many other useful tools available for all aspects of project and resource management and, if you are managing a portfolio of significant projects, you should ensure you are aware of what is available and how it might help within your organization. Ensure also that you acquire proper documentation and, if required, training for the use of such tools. No matter which tools you use, you will find the BANTAM methodology extremely valuable in supporting the overall management of programs and projects within your organization.

5.4 Training

Training is naturally important in relation to any new operational application or initiative. However, in the context of a biometric or token technology application it is especially important, as we are dealing with technologies which require a particularly personal interaction between system and user. This is especially so for biometric applications where users are providing a physiological sample of themselves at each transaction instance. Because of this very personal manner of interaction with the technology, there are bound to be some strong views, and possibly concerns, which need to be addressed in addition to the obvious usage training which you would expect with any system. Being able to explain and demonstrate exactly how the system works is vitally important if we are to properly address such issues, and this includes areas such as infrastructure and architecture. The BANTAM maps created during application and process definition will prove invaluable in this respect, as we will be able to show precisely how the application has been designed and what is happening to user related data within operational systems processes and transactions.

First of all, we need to train the trainers and ensure that all administration staff who will be responsible for the day-to-day running of the system, have a thorough understanding of the application and how it functions. Those who are going to be responsible for enrolling others into the system, also need to be aware of the people related issues around this, and how to generate high quality reference templates when biometrics are used. In the case of tokens, these individuals also need to thoroughly understand the processes and procedures around issuance, accidental loss and replacement, as well as the technicalities of encoding them properly in the first instance. Again, they need to know exactly how the application is working and what is happening to the data provided by the user or tokens in the user's possession, at all times, during enrollment and subsequent transactions.

In addition to the above, we naturally need to consider a program, or series of programs, for training users of our new application. The complexity of such a program will naturally depend upon the scope and sophistication of the application, but in all cases we shall need to consider training for both operational processes and actual interaction with the system. Fortunately, we have a wealth of pertinent information at our disposal if we have been using BANTAM. What we need to do now, is to capture the relevant information for use within carefully considered training modules for both administration personnel and end users.

BANTAM provides a simple, yet effective framework to help you create a training program consisting of several modules. The delivery of these modules may take a variety of forms, including computer based multimedia, classroom presentations, written instruction, actual system demonstrations, or maybe a combination of these techniques. In any event, we shall need to coordinate and publish the training modules in a consistent and logical manner. We shall also need to set up a log of who has been trained in what discipline and, if there are large numbers of users involved, put together a program for rolling out training in the most efficient manner. Finally, depending upon the scope of the project overall, we may wish to provide a formal certification for administration personnel undertaking training, in order that they may be recognized, and able to operate throughout the organization in the context of managing elements of the application.

The BANTAM documentation takes the form of a Training Schedule, within which there may be several training modules. You may use multiple training schedules in order to differentiate between users and administration personnel.

Schedule ref.		Schedule Title	
Date of Issue		Process	
Prepared by		Certification	

Bantam Maps Included in Training Schedule	
Document Title	Document Reference

Figure 5.12 The BANTAM Training Schedule header

The Training Schedule header, as shown above, sets out the pertinent details of the schedule, starting with a unique schedule reference and a schedule title. The date of issue field shows the currency of this document and the process field describes the process for which this training schedule has been developed. Lastly, the certification field identifies any formal certification which may be issued after successfully completing the training. The BANTAM maps summary provides details of the BANTAM maps used with this training schedule and acts as a checklist for trainers and students alike.

The Training Schedule continues with the following standard sections which must be completed for each schedule.

- Schedule Overview. The schedule overview provides a plain language description of the training schedule and the areas covered within it. It will also outline clearly the purpose of the schedule.

- Target Audience. This section sets out the target audience for this particular training schedule. It may include details of role, department, location and so on, depending upon the scope of the project and the user population in question.

- Required Hardware. In this section, details will be provided of any hardware, such as computers (including minimum specifications where appropriate), printers, capture devices, token readers or other such equipment which will be necessary in order to undertake the various training modules within this schedule. Where appropriate, manufacturer's details and model numbers should be quoted.

- Required Software. The required software section describes any software required for training purposes, including specific operating systems, supporting software components and the main application software (including version numbers where applicable), as necessary to undertake this training schedule successfully. This is an important section, as, when software is updated to a new version, perhaps to accommodate enhanced functionality, then, naturally, the training schedule and associated materials will also need to be updated.

The above sections are provided mostly for the tutors who will be delivering the training modules and therefore need to understand both the scope of the training and the tools required to run the training modules successfully. It is important that such details are clearly defined and adhered to if we are to provide consistent, high quality training throughout the program. Similarly, when new personnel are subsequently required to use the system, they should receive the same level of training, provided under the same conditions, as the original user population. Note, that this may be months or years after the date of the original implementation, making the need for clear and consistent training documentation, as provided by BANTAM, especially important. The value of adopting such an approach will become particularly apparent when the system functionality changes and it is necessary to retrain those who were trained on the previous version of the system, possibly using different hardware or software. The next section of this document states the number of modules within the training schedule and provides a summary checklist accordingly.

Number of Modules within this Schedule	

Training modules included within this Training Schedule	
Module Number	Module Name

Figure 5.13 The module checklist

Following on from the information described above, and attached to the main document, are summary sheets for each module within the overall training schedule. These individual module sheets contain the module number, module name and schedule reference as a header, together with a Module Tutorial section, which provides the written tutorial itself. In some cases, this may take the form of several pages, together with illustrations. In other cases, it may reference other material such as multimedia presentations and, of course, the relevant BANTAM maps, as well as actual demonstrations given by the tutor.

Having a BANTAM Training Schedule for each significant element of the application ensures that the application will be used as intended and that users are properly trained and given pertinent information about the application and how it operates. It will also provide the opportunity for ongoing feedback as users make their own observations during the training program. This feedback may be extremely useful as you fine-tune the application and consider future enhancements. An additional benefit will be realized, in acknowledging the importance of proper administration and the position of administration staff accordingly. If administration personnel undertake a formal and vigorous training program, the benefits to the subsequent performance of the new application will be significant, as users will be properly enrolled and instructed on how to undertake a successful transaction. In turn, this will result in fewer 'help desk' inquiries.

We have emphasized the training element quite heavily, but it *is* extremely important. Remember, we are dealing with relatively new technologies here, which, while they may be familiar to those directly involved as manufacturers and vendors, their applications are still far from being familiar territory as far as end users are concerned. In addition, the introduction of technologies such as biometrics and chip cards into mainstream operations very often results in significant changes to existing processes as well as totally new operational processes. In many cases, we are indeed introducing a new culture into the organization or one of its key processes. Compare this with the introduction of any other 'culture' change within your organization. Would you expect that any such significant change would be introduced without a proper training program? Probably not. Furthermore, it would be common to assign suitable individuals as 'trainers' in order to be able to explain the new processes, answer questions and provide on-the-job advice where required. Similarly, with our new biometric or token technology application, it is important to cover these training and communication issues properly and ensure that we have sufficient numbers of trained individuals on hand to cover the introduction of the new technology and ensure that all goes smoothly. In the past, this hasn't always happened, as all too often such applications have been shuffled under the 'security' banner and not given the proper emphasis and resources in order to do the job properly. This usually proves to be a short-sighted approach, often costing heavily in subsequently wasted time and energy while the application is bedded in. Sometimes, it has resulted in perfectly good technology being withdrawn, simply due to ill-considered implementation and poor training.

BANTAM has provided a comprehensive tool set with which to ensure that the application is properly conceived, designed, purchased and implemented, but the job doesn't stop there. If we are to enjoy a successful application, we must pay attention to communication and training at all levels.

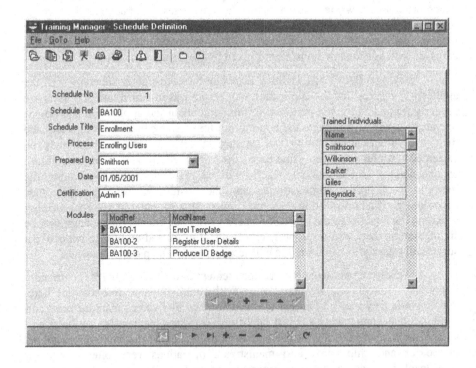

Figure 5.14 The Training Manager module of BANTAM Program Manager

The Training Manager module within the BANTAM Program Manager allows you to define and log all of your required Training Schedules, together with the individual modules which make up the schedules. It also allows you to assign individuals to specific Training Schedules, as considered necessary within your particular organization, and to print out reports of those who have undertaken the training and is therefore qualified accordingly. You could do this within another application, or build a separate database to track such information, however, the BANTAM Program Manager does provide a convenient mechanism to do this, and integrates the function with related document management and much more. There is a trial version of the program on the accompanying CD-ROM which you may evaluate yourself.

We have covered quite a lot of ground as to the basics of the BANTAM methodology and how it may be used to good effect in the contexts of biometric, token technology and related projects. We have also covered the associated documentation and how BANTAM may be used in areas of procurement and overall program management. Finally, we have covered the important area of training and stressed that this must be properly considered and addressed if we are to realize a successful application. Hopefully, by now, the reader has a pretty good idea of what BANTAM is all about and how it may be used to great benefit within various related projects. The next section of this book will take the form of a working example, in order to give a feel for how BANTAM might actually be used in the context of a real life project. Naturally, the project is fictitious, but perhaps not too dissimilar from those currently being considered in many government departments, corporations and other organizations around the world. Your own organization may of course be quite different from that depicted, but nevertheless, there will doubtless be similarities in some of the issues addressed.

6. A working example

In this chapter we shall explore the use of BANTAM within the context of a typical requirement as may be considered by any organization. For our purposes, we shall use a fictitious organization which we shall call International Financial Management Inc. or IFM.

IFM's requirement is that they wish to provide secure access to their main network from anywhere in the world for their 1700 employees and trusted agents. This is in order that they be able to communicate confidential sales and marketing information, accept the submission of bids or sales proposals, receive and submit individual monthly reports and that the sales agents may access new product illustrations from anywhere, even if they are on client premises. This obviously entails the use of untrusted networks such as public switched telephone networks and the Internet. Their main concerns are twofold. Firstly, that they have a high level of confidence as to the true identity of the individual being granted access to the corporate network and, secondly, that information being passed between the remote client and the network is secure during the session. They have placed one constraint on their staff in that the client device used for access must be one of their own specially equipped notebook computers. Third party devices must not be used.

The systems support team at IFM have a rough idea of how the application might be configured. They wish to initiate virtual private network (VPN) sessions with their remotely sited representative using a biometric for identity verification purposes. They are open-minded about the possible use of digital certificates, although in general they doubt that this will be necessary. They envisage the use of a specially equipped notebook computer which will have a standard build of application software, including the client VPN software and biometric functionality in addition to specific business software. Customization of this standard build software will not be permitted and staff will be actively discouraged from loading any other applications on to these notebook devices. Within the boundaries of the corporate network, they will maintain an access control hierarchy in order to determine individual access rights. In addition, they will log every instance of remote connection in order to provide a comprehensive audit trail.

The IFM management have sanctioned a program to investigate the options and, if applicable, instigate a phased roll-out of the new connectivity model to all of their international staff and agents. An experienced program manager from the sales and marketing area, John Getz, has been appointed to organize and run this particular initiative on behalf of the company. John will liaise with both internal and external consultants as appropriate in order to understand the bigger

picture and evaluate the feasibility of implementing an application of this nature.
John has decided to use the BANTAM methodology throughout, to ensure that the
various issues and requirements are understood and documented properly. He has
also decided to use the BANTAM Program Manager software as a personal aid to
gather together relevant documentation and personnel details. If the planned
solution turns out to be feasible from both a technical and cost perspective, then
John will configure a series of mini projects within the overall program in order to
phase the implementation on an area-by-area basis. However, for the purpose of
our illustration, we shall concentrate on just one 'project'.

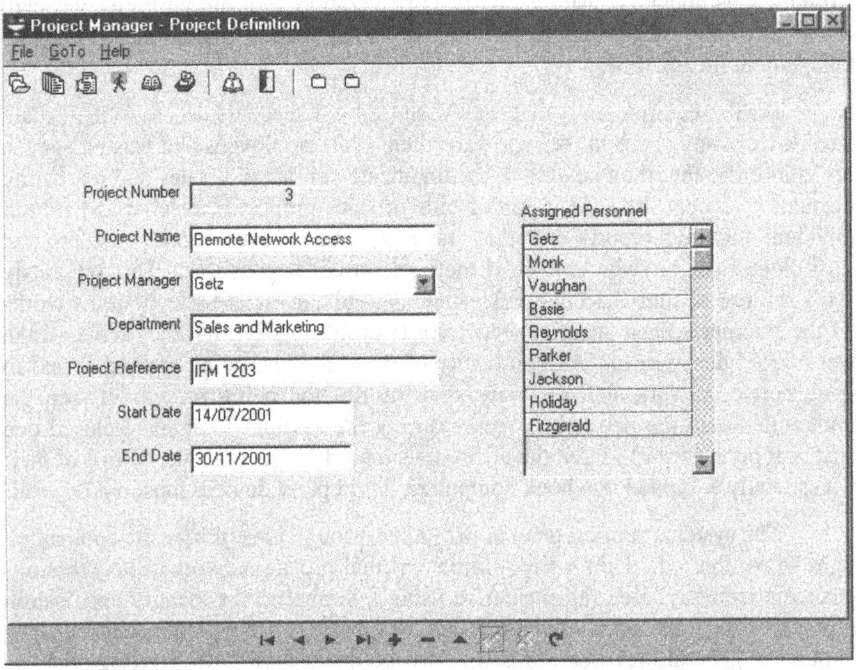

Figure 6.1 Setting up the project

The first thing John does is gather together his project team and set up the
project in the Project Manager section of the BANTAM Program Manager, which
will help him produce reports and keep track of documentation appertaining to
specific projects. In fact, the BANTAM Program Manager may be used in this way
for any project or group of projects. He will also start to identify potential
suppliers and external consultants and enter their details into the Supplier Manager
section of the BANTAM Program Manager. Internal staff are entered into the
Personnel Manager and the Document Manager will be used to log all project
related documentation. As a starting point, John would like to see an Application

Logic Map, to serve as an initial discussion document. Billie Jackson is the IFM Systems Manager as well as part of the project team, and has been tasked with producing this document accordingly. Billie will liaise with other personnel as necessary in order to capture the relevant information.

6.1 Articulating the requirement

As previously stated, IFM have a good general idea of the requirement, together with some ideas on how it may be met. However, this is not enough if we are to progress towards an intelligently conceived application. We must be specific about our requirements and the associated benefits being sought. Billie Jackson thus has the task of producing the high level requirements definition as a first draft BANTAM Application Logic Map document. This document may develop through iterations as a result of subsequent discussion until a final version is produced and signed off by the project manager.

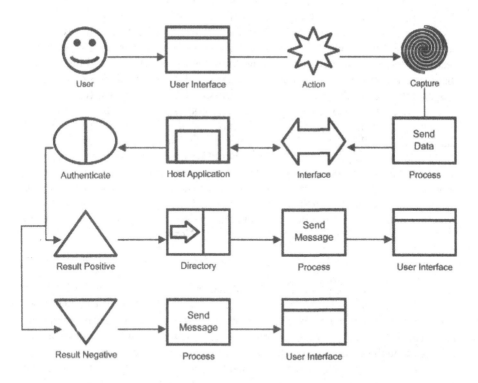

Figure 6.2 The Application Logic Map graphic

The graphic for the first iteration of the Application Logic Map as depicted in Figure 6.2, is deliberately kept simple and just detailed enough to act as a general discussion document. The accompanying Map Explanatory Notes explain that, via a user interface, biometric data is captured and sent via an interface to the host application where an authentication process is undertaken. Upon a positive result, the corporate directory is referenced in order to ascertain access privileges and a message is returned via the user interface accordingly. Upon a negative result, a message is simply returned to the user to indicate that occurrence. The Map Explanatory Notes also indicate that a custom notebook computer device is anticipated as the client, and that this device will be loaded with specific application and communications software in order to enable the required functionality.

Now that this top level statement of requirements has been produced and circulated, John Getz will call together the project team for the first time in order to discuss the requirement in more detail and listen to the various questions and concerns, both from a technical and practical point of view. At this inaugural meeting, the current, unsatisfactory method of remotely accessing the network is discussed. The current technique involves the use of random number generating tokens and passwords, which are considered over-complicated and fussy. In addition, the functionality currently offered is heavily restricted in relation to the relatively low confidence level as to the true identity of the user. While it is felt that when the remote device (notebook PC) is under the jurisdiction of the *bona fide* user there is a relatively low risk of access by the wrong person (intentionally or otherwise), it is also felt that should the remote device be stolen or otherwise misappropriated then there is a good chance that the token may be with it and therefore offer little security. Given the current view and operational procedures, it is decided that a Logical Scenario Map should be produced in order to document the user experience, and that a number of Functional Scenario Maps would be produced as necessary in order to capture the various functional areas of the application. This task is allocated to the analysts Gary Vaughan and Martin Basie, who will liaise with their colleagues as necessary in order to produce the top level BANTAM maps and associated Map Explanatory Notes.

In parallel with these activities, John Getz has been researching the market place and building up his database of suppliers and external consultants accordingly. By using the Supplier Manager module within the BANTAM Program Manager, he is able to collate all of this information in one place and carry it with him on his notebook computer, wherever he may be working at the time. This is extremely pertinent in this case as John will be visiting various branch offices throughout the course of this project and will need to be able to work effectively from any one of them. In addition, when connected to a modem, he may call any contact from any supplier, on either a switchboard number or mobile, straight from the database with a single button click This also applies to contacts listed within the personnel manager database. The document information held within the BANTAM Program Manager will prove invaluable, as will the potential to hold relative data for an unlimited number of projects. Another useful piece of functionality is the ability to produce, quickly and easily, a number of standard reports appertaining to a given project and print these to any local or network printer. Indeed, for John, using the BANTAM Program Manager is a little like having a super-efficient private secretary with him at all times.

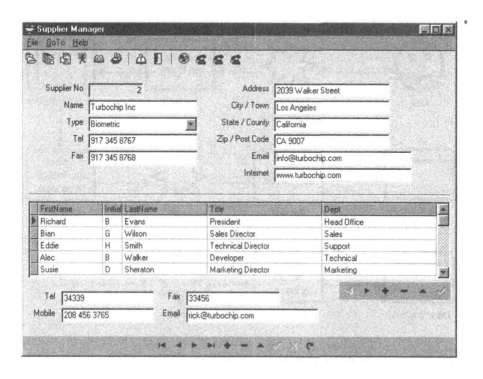

Figure 6.3 The BANTAM Program Manager Supplier Manager module

However, back to the logical scenario map being prepared by the analysts. Before capturing a suitable process, Gary Vaughan undertook some research among current users in order to understand their perception of biometric technology and how they might wish to use it. He found an overwhelmingly positive attitude among IFM's employees and agents, all of whom were keen to try the new technology. In addition, several users suggested ideas for how the technology might be implemented, out of which came an obvious requirement for protecting access to the notebook computers, whether or not they were connected to the network. The point was also made that, in the unlikely event of a notebook PC being stolen, it would be harder to access data if both the log on and individual application access were to be protected in some way. Gary asked users whether they would be prepared to give their biometric multiple times in order to access specific applications or the corporate network, and he found that the vast majority would happily agree to this. Proper research among prospective users of an application is always a sensible thing to do. Often, we may be pleasantly surprised by the informed attitude of users and the general willingness to participate in initiatives aimed at providing better services. This was certainly the case with the research undertaken by Gary Vaughan, which showed strong support for the initiative under consideration.

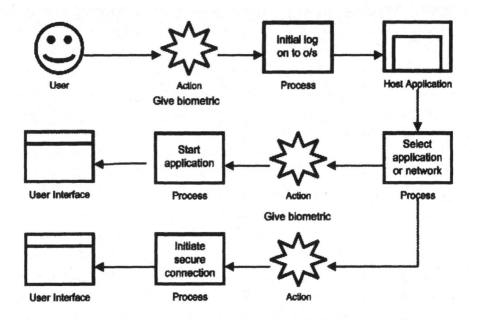

Figure 6.4 High level Logical Scenario Map graphic

From the user's perspective, it was ascertained that the high level logical scenario depicted in Figure 6.4 would seem intuitive and completely acceptable in terms of an authentication process to enable access both to the local machine and the corporate network. In this instance, users would be required to give their biometric during the operating system boot sequence, ensuring that only the *bona fide* user could start the PC in the first place. The notebook PCs used will have a special software build with limited user access to operating system features and a common set of operational applications. Having logged on to the operating system, the user may choose to start an application residing on the local PC, in which case, if it is classed as an operational application, they will be required to give their biometric again in order to launch the application. They may then use the application as required. If the machine is unattended for a period of time and the screen saver activates, then the user must give their biometric again in order to unlock the machine. If an active application requires data that is residing on the corporate network, then the user must give their biometric again in order to initiate a secure session between client and server as appropriate. Similarly, if the user wishes to navigate immediately to the corporate network in order to access an application or specific data, then the biometric must be given again. In the case of the IFM users, this multiple stage verification was actually preferred over the option of a single sign-on approach from machine start up. Users felt that it was no inconvenience, and quite acceptable, to provide their biometric in order to access deeper levels of functionality or company specific information.

Having captured this simple, high level logical view from the users perspective, it was now considered pertinent to produce Functional Scenario Maps in order to provide a little more detail as to how this functionality would be provided. It was decided to produce two maps initially, one for local access and one for network access. Gary Vaughan and Martin Basie would work together to produce these maps.

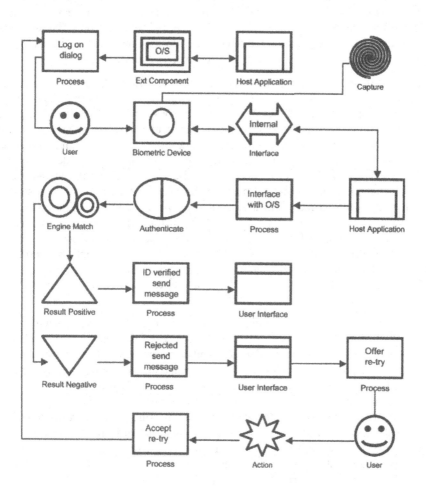

Figure 6.5 Initial log on Functional Scenario Map graphic

In Figure 6.5, the initial log on function is depicted at high level. In this instance, as part of the boot up process, the operating system presents a dialog to the user which has been modified by the host biometric application to include instructions to the user to present their biometric. Upon doing so, the biometric sensor captures the relevant biometric data and interfaces with the host application in order to provide this data to the application. The interface is shown as internal as we are anticipating notebook PCs with integral biometric sensors as the remote clients. The host biometric application interfaces directly with the operating system security model and an authentication process is initiated via the on-board template matching engine, which results in either a positive or negative matching result. If the result is positive and the user's identity has been verified, then a message is sent to the user interface accordingly and access to the operating system is granted. If the result is negative, then access to the operating system is denied, a message is sent to the user interface to this effect, and a retry process is initiated which offers the user a defined number of retry attempts to log on. The user then takes a definite action to accept the retry process, which results in the initial log on dialog being re-presented to the user, and the process starts again.

This particular Functional Scenario Map is still presented at a relatively high level and it may be necessary to produce a lower level map in order to show more detail around parts of this functionality. For example, the precise interface and interaction between the host biometric application and the operating system may need to be explained in more detail in order for IFM's technicians to be able to approve the final application. This level of detail may be provided at a later stage when the IFM project team are in discussion with preferred suppliers. Similarly, detail around exactly where and how the biometric template is stored, and how it relates to the client user directory, will need to be understood. The supplier might usefully employ BANTAM to illustrate this concept. However, at this stage in the project, the Functional Scenario Map graphic depicted in Figure 6.5 will be sufficient to show, as one of a series, to potential suppliers in order to illustrate how IFM is thinking about the application. Subsequent discussion will clarify the lower level detail and additional Functional Scenario Maps will be produced as considered necessary for this project.

Important among this initial series of Functional Scenario Maps will be the map which describes the anticipated network access functionality. Martin Basie has been considering the options and has had discussions with both potential users and also the IFM network administrator team who, naturally, are very interested in this application and the benefits it may bring to secure network access in general. It is anticipated that biometric identity verification will take place at the client position and will be closely integrated with the operating system security model as previously suggested. This cuts down on potential network traffic and simplifies the central directory and related infrastructure. The resulting security model is considered acceptable for the requirement at hand at this time. However, a centrally verified biometric access model may also be considered, allowing an alternative methodology should the initially chosen approach prove to be inadequate.

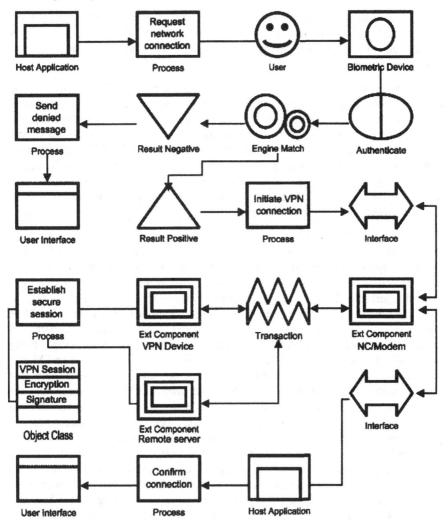

Figure 6.6 Remote network access Functional Scenario Map graphic

The Functional Scenario Map depicted in Figure 6.6 is, once again, fairly high level in this first iteration and intended more as a discussion document than a definitive statement of how the application will function. However, it does provide a good starting point and something for the application developers to build upon. It shows a remote network access request from the host application resulting in a biometric authentication process on the client machine, which, if successful, will result in a virtual private network (VPN) link being established. The object class symbol, appended to the secure session process symbol, shows us that we are anticipating strong encryption and digital signatures as features of this session.

The map also shows bi-directional communication between the VPN device, server and network device or modem at the client end. If a secure connection is established, a message is sent to the user interface to confirm the fact, and a session may begin.

However, while this particular map covers the initial connection to the network, there is still the question of access control within the network to consider. Exactly which network domains and applications are accessible to a particular user will need to be configured in the usual manner via access control lists. This route is considered acceptable to IFM for now, as they will have a high level of confidence as to the true identity of the individual establishing the connection, and will be using strong encryption for all traffic between client and host via a VPN session. The VPN device details, firewalls and server configuration will be detailed in a separate BANTAM map.

It occurs to John Getz that, at this stage, they should really have a high level view of their current systems architecture, at least as it might affect this particular application. He therefore requests that a suitable Systems Architecture Map be produced by the analysts working on the project, in order that this be viewed together with the various Logical and Functional Scenario Maps in order to have a complete high level picture of the proposed application and how it will be integrated into the existing infrastructure. This will make it easier for the technical team to understand the configuration and integration possibilities for what is being proposed. In order to produce such a map, the analysts Gary Vaughan and Martin Basie considered it best to start with the existing documentation and see how that aligned with the current reality. In doing so, they quickly discovered that the scanty documentation that was available was seriously out of date and did not reflect the current position. In addition, the accompanying notes were not particularly clear and did not explain how decisions had been arrived at, or hint at any ongoing strategy. They decided then and there that, from now on, they would use the BANTAM methodology to describe architectural areas within the enterprise, both graphically and in descriptive text. Furthermore, such documents would be subject to version control and regular reviews, using BANTAM as a framework within which to accomplish this. There would be an associated benefit in that any future projects requiring such supportive information would have access to properly prepared BANTAM maps which they could then use directly within their own projects, thus facilitating reuse and the associated efficiencies that this way of working provides. Furthermore, it would ensure that any agreed changes due to other program initiatives would be fed back to the prevailing architectural view, enabling the document set to be continually revised and therefore kept current. When adopting such a course, it is likely that you will find, as Gary and Martin have, strong support from other areas within the enterprise who would also like to benefit from this holistic way of working and sharing information. It is an example of how the scope of BANTAM may be broadened to provide benefits throughout the organization, and for a variety of related purposes. Furthermore, the wider it is used, the greater the benefit realized in doing so. A true win-win situation.

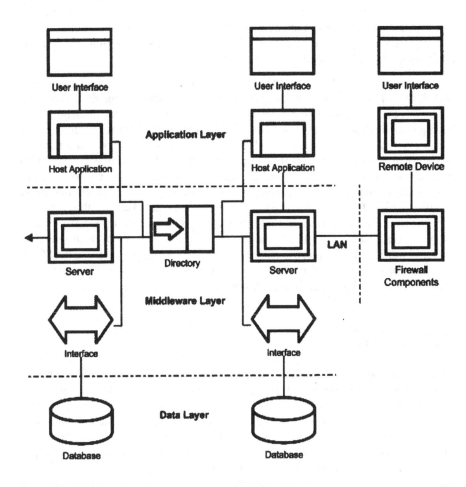

Figure 6.7 Conceptual systems architecture

Figure 6.7 above gives a conceptual view of the systems architecture at IFM and depicts a multilayer approach to separate out data sources, operational application logic and the client application/user interface areas. It also shows a common directory sitting on the LAN and being accessed as required by the application and middleware layers, and a separate firewall layer allowing remote access to the LAN. This high level conceptual view will be complemented by lower level maps showing precise details of domains, server identities, interfaces and communications protocols for a given geographic or operational area. In the case of IFM, they will produce a series of BANTAM maps depicting the architecture from different perspectives and using the reference symbol accordingly. Using the BANTAM Program Manager software, they will set up a perpetual project named 'IFM Architecture' and collate all of the associated maps together in one place, in order that they be easily managed, viewed and updated accordingly. Using the Report Manager, they will also be able to produce a current listing of all such documents at any point in time, with a single button click. This will represent a significant step forward from their current position, but will be easily managed with a minimum of resources.

The BANTAM maps we have looked at so far have been relatively simple, high level affairs which have taken little time to produce using the BANTAM methodology. However, you can see how valuable they have been already, just in having something documented as a basis for discussion and further development of the original idea. We shall assume that IFM have produced a few other maps at this level in order to capture pertinent elements of the application as perceived by themselves at this stage. They can now approach potential suppliers with a clear idea of how they see the requirement, together with a possible configuration solution which fits with their particular environment. Having this level of understanding and documentation at this stage, will save both IFM and potential suppliers significant amounts of time as they explore the options with which to move forward. Furthermore, the well researched maps already produced may form the basis for the next level of detailed design, without having to start from point zero. In BANTAM, nothing is ever wasted.

6.2 Issuing an RFP

IFM have already issued a series of Requests for Information (RFI) and undertaken their own market research, and so, have a pretty good idea of who is supplying what within the technology areas of interest. They have produced a series of high level BANTAM maps, which have proved extremely useful for internal discussion among the relevant parties, and now have a clear view of their requirements and how they expect them to be met. They are now ready to approach external suppliers and issue a formal Request for Proposal (RFP), for which they will use the BANTAM provided template. Issuing RFPs and evaluating the responses in a consistent and objective manner is a vitally important stage within your overall program. It is important that this activity is properly planned and orchestrated as part of the project, and that sufficient time and resources are allocated accordingly. An incorrect decision at this stage could have serious implications further down the line, so we must ensure that we carefully consider all available options and fully understand their relevance and suitability to our particular situation. Fortunately, IFM have prepared well for this stage with much internal discussion and, of course, a comprehensive set of BANTAM maps with which to articulate their requirements. Prospective suppliers will therefore be in no doubt as to what is being asked of them, or how they should respond. All they need to do now is to attend to the mechanics of issuing the RFPs and set some realistic dates for receiving back proposals and evaluating them accordingly. The recipients have been chosen, so we now need to prepare a suitable RFP document.

The BANTAM RFP provides clarity and consistency with a summary of all the pertinent points. It is issued complete with a set of BANTAM maps in order to illustrate the requirement clearly, and additionally provides a standard form of response to facilitate easy evaluation and comparison of the proposals received. The first part of this document sets out the supplying organization's contact details and provides a summary of the attached documents.

Bantam | Request for Proposal

Respond to			
		Date of issue	14/07/2001
Organisation	International Financial Mngmt	Response req. by	14/08/2001
Name	John A Getz	Reference	RFP 1003
Title	Program Manager		
Address	1327 Pharaoh Drive		
Town	Alexandria		
County	Virginia	Telephone	916 345 786
Post Code	VA 2609	Fax	916 345 787
Country	USA	E-mail	getz@ifm.com

Project Name	Remote Network Access

Bantam Maps Included in RFP	
Document Title	Document Reference
High level systems architecture	SAM 001
Specific authentication architecture	SAM 002
High level application depiction	ALM 001
Client log on map	FSM 001
Network access map	FSM 002
Current process map	LSM 001

Figure 6.8 The BANTAM RFP header

From this, the respondent can see exactly who they are dealing with, how to contact them, the date a response is required by and which documents are attached to the RFP. The project name is also included, as it is feasible that a supplier may receive more than one RFP from the same issuing organization. It also helps when contact is made, especially if the issuing organization happens to be running several projects at that time. Notice that the BANTAM RFP places this information right at the start of the document, where it is easily found – respondents don't have to go wading through a huge document in order to find who they are dealing with or who to call. The next section of the document provides a box for a plain language textual description of the project. You may think of this as a sort of 'Executive Summary' which gets straight to the point and clearly outlines the requirements of the issuing organization. After reading this section, the respondent should be in little or no doubt as to their capacity to meet the requirement, at least in broad terms. If they feel that they can meet the requirement, then they may choose to continue through the rest of the document. Even at this high level, we can complement the textual overview with a BANTAM map in order to show the overall requirement graphically. In this case, the applicable map reference may be shown under the text box. In fact, every main section of the RFP has a map reference bar underneath it.

Project Overview

> The purpose of this project is to provide secure remote network access for home workers and those travelling away from the main office buildings. The solution should be easy to use from a users perspective, secure and easily maintained. Furthermore, it should be scaleable and provide flexibility for subsequent development at a later stage.
>
> It is anticipated that users will use a notebook computer with integral biometric fingerprint sensor in order to verify their identity when initiating a session between remote client and network. These devices will be rolled out on an area by area basis, together with the application currently under consideration, following a pilot implementation at the head office location.

Map Refs	ALM 001				

Figure 6.9 The RFP project overview section

There now follows a whole page dedicated to a textual description of the detailed requirements of the project. This is where we can describe, in as much detail as necessary, for what exactly we wish the respondent to propose a solution. In this section, we shall reference the attached BANTAM maps quite extensively, ensuring that the respondent has a detailed and full picture, both of our requirement and the environment in which we wish to implement the application being sought. It is the provision of this sort of detail that makes the respondent's life a lot easier, with the clear articulation of your requirements supported by relevant high quality documentation. IFM have paid particular attention to this area and will reap the benefits accordingly, in the form of intelligently conceived proposals which closely match their requirements and take into account their particular infrastructure.

A separate box is provided within the RFP in order to capture any special operational or architectural conditions which may apply and need to be highlighted. This is useful, for example, if you have irregular network traffic or perhaps an unusual technical infrastructure in a certain location or, indeed, any condition which you believe the respondent should know about and take into consideration while preparing their proposal. In the case of the IFM project, it has been noted that their level of network traffic is variable and typically increases towards the month end when reports are being prepared and much data is being shared in preparation. It is important to IFM that this situation does not severely impact upon the ease with which personnel can log on to the network and obtain the information they need. This, in turn, highlights the need to look at network capacity carefully and anticipate both current and future requirements.

Special Operational / Architectural Conditions

Performance must be independent of remote location or time of day and be scaleable for up to 1700 users. Use of public communication networks should take into account the variability of bandwidth in certain locations. Many users will be mobile and will rely on conventional dial up modems for connection purposes. IFM has heavier network traffic towards month end periods when reports are being generated and shared across networks.

Map Refs	SAM 003				

Figure 6.10 The special operational/architectural conditions box

Having noted any special conditions which may apply in the case of this project, it is reasonable to assume that, depending upon the precise nature of the requirement, potential suppliers may wish to visit the issuing organization in order to confirm their appreciation of the situation and ask any relevant questions. If this is to be allowed, then it should be properly organized on a basis which is consistent and fair to all concerned. Naturally, the BANTAM RFP provides for this also, with a special section for site visits and associated information.

Site Visits

Available Yes / No	Yes	Preferred Days	Tuesdays
Contact Name	Carmen Fitzgerald	Preferred Time	11.00 – 15.00
Department	Sales and Marketing	Notice Req.	5 days
Title	Executive Secretary	No. Of Visitors	2
Telephone	916 345 8876	On Site Parking	Yes
E-mail	fitzgerald@ifm.com		

Related On Site Requirements

Visitors must register with security upon arrival. This is a strictly non smoking site. Parking is only permitted in designated areas and visitors parking permits (available from security) must be displayed on all vehicles. Visitors must read the fire and emergency information notices available at reception.

Figure 6.11 Site visit related information

This straightforward dialog, as depicted in Figure 6.11, once again gets straight to the point and advises on what basis site visits will be allowed and who to contact in order to arrange a visit. It also advises on car parking procedures and other related on site requirements. When vendors become increasingly familiar with the BANTAM RFP, they will appreciate that this sort of concise information, presented in a reliable and repeatable manner, will prove invaluable and will save much time. In addition, it dispenses with ambiguity and allows the issuing organization to follow a clear and consistent policy with regard to site visits. This section of the RFP also includes a box where the issuing authority may provide some additional information about their organization. This again is valuable, as both parties need to know who they are dealing with. Furthermore, providing such information will help the potential supplier to understand the context of the requirement and tailor their response accordingly.

The next section of the RFP states clearly the format of response required, the procedure for questions and how the responses will be evaluated. It is important to be clear on these issues, for the benefit of all concerned, and especially in order to ensure that responses may be easily compared.

Format of Response Required

> Respondents must complete the attached BANTAM RFP Response Summary and provide their own series of BANTAM maps in order to describe their proposal in line with this RFP

Procedure for Questions

> Contact John Getz on 5516 340 9989 only. No other contact will be permitted and approaches to other IFM personnel may result in disqualification of any proposal submitted. No further questions will be entertained after 12/08/2001, at which point all responses should be received.

How we shall Evaluate Responses

> We shall compare properly formulated responses on a like for like basis, in line with our stated requirements. The quality of technical solution, coupled to feasibility and ease of implementation shall take precedence. No communication, unless initiated by IFM, shall take place during the evaluation period. Respondents shall be notified in writing as to the outcome.

Figure 6.12 Understanding response requirements

Once again, this is fair to both sides and removes any ambiguity or assumptions. Potential suppliers know exactly what is required of them in the form of a response, who to contact if they have any related questions, and how the responses will be evaluated. Together with the quality information supplied within the BANTAM maps, they now have a complete picture of the situation and how they should respond if they are interested in submitting a proposal. Naturally, the issuing organization must adhere rigorously to these conditions themselves, in order to ensure consistency.

The last section in this part of the RFP is the signatory box. This states clearly who within the issuing organization has sanctioned this RFP, their department, their position within the organization and the date on which the RFP was approved and signed, plus, of course, their signature. This is a small but important section of the document. From the supplier's perspective, it confirms that this is an official request and that it will be progressed and managed accordingly. From the issuing organization's perspective, it provides an audit trail of exactly who has raised such RFP documents. This, in turn, encourages them to develop and maintain a clear policy on who has the authority to issue and sign such documents. Bear in mind that in some cases, we may be dealing with projects on a significant scale, requiring equally significant financial commitment from the organization concerned. We therefore need to be clear on this point, and understand the internal procedures leading to the creation and approval of an RFP. It would also be unfair to potential suppliers if such documents were issued without such a procedure.

With the BANTAM RFP document, we now have a vehicle with which to address the key points of any RFP effectively, and to support these key points with qualified documentation in as much detail as is necessary to describe our requirements. Furthermore, the use of the BANTAM RFP promotes clarity and consistency in the way in which your organization communicates with the outside world in relation to products and services required.

6.3 Evaluating proposals

Having issued a well considered RFP, we now need to consider how we are going to evaluate the responses received. It will help if all potential suppliers respond in the same way, and BANTAM makes this easy for them with the provision of a standard response summary. This is in fact part of the original RFP document, and deliberately so, as we need to be able to tie responses to original requests, in many cases after the event. Responding organizations thus return a copy of the original RFP, with the response summary duly completed as appropriate. This in itself ensures consistency and makes it easy for the issuing organization to compare the high level information associated with each response. Lower level detail, including BANTAM maps produced by the potential supplier, will be appended as necessary and referenced within the response summary, together with any other documentation provided. The issuing organization should study the information received within each response summary carefully and eliminate any proposals which clearly do not align with their requirements. There should be a defined procedure for undertaking this task. In the case of IFM, John Getz has organized a special workshop, where representatives from both technical support and business areas will be present. Under the chairmanship of Getz, this group will open and evaluate the summaries of all responses received by the appointed date. Those which meet the criteria for serious consideration will progress further and will be analyzed in detail by the project team following this workshop. For any proposals which are considered unsuitable at this stage, either for reasons of cost or inexperience on the part of the responding organization, a standard letter will be drawn up, explaining this fact and offering a brief explanation of why such a proposal will not be progressed further. In fact, this would be an extremely rare occurrence when BANTAM is used, as potential suppliers will be in doubt as to the precise requirements of the issuing organization, and therefore, their own ability to meet these requirements. Obviously, it would be in nobody's interest to waste time preparing a proposal which clearly doesn't satisfy the requirement. It is anticipated then, that, in the majority of cases, all proposals received will be acknowledged and progressed to the next stage of evaluation.

The Response Summary provides a concise overview of the responding organization, their related experience in this field, and of course the estimated cost of their proposal. It also clearly sets out contact information and responsibilities within the organization concerned.

Response Summary

About the Respondent

Organisation	TurboChip Inc.	Core Business	Device manufacturer
Principal Contact	William Fields	Years in Business	9
Title	VP Technology	Annual Turnover	$23m
Address	1327 Ocean Way	Registered No	1099347
Town	Santa Monica		
County	California	Telephone	315 657 3329
Post Code	CA 9007	Fax	315 657 3330
Country	USA	E-mail	fields@turbochip.com

Principal Company Officers

Name	Title	Telephone
Jack Smetana	President	315 657 3327
Randolph Grieg	Chief Financial Officer	315 657 3325

Other Pertinent Information

TurboChip have undertaken several similar projects in recent years, some of which have been on a broader scale than that being proposed. We are capable of providing a turnkey solution, including all required consultancy, application design and implementation. We offer ongoing support for all solutions provided by ourselves.

Figure 6.13 Response Summary header

The Response Summary header identifies the principal contacts, how long the company has been in business, their annual turnover and other details. It also provides the responding organization with the opportunity to elaborate on this a little and provide other pertinent information. So, now we know who we are dealing with. We also know the names of the principal company officers, in case we wish to make contact in order to qualify any of the information provided. Having established that this organization seems to be qualified to offer such services as we require, we would rather like to see some evidence of previous experience, including, where applicable, a reference site with whom we may make contact at our discretion.   Many suppliers of course positively welcome the opportunity to provide a reference site, and this in itself may be a good sign, depending on who exactly the reference is, and how similar the application provided.

Similar Projects Undertaken

Client	Description	Date
Bison Foods	Remote network access for 1200 users	Jan 1998
Monks Bank	Internal secure access for dealing purposes	March 1999

Reference Site Details

Organisation	Nelson and Green	Contact Name	Billie Klemperer
Address	1595 Wilson Street	Telephone	415 347 2289
Town	Chicago	E-mail	Billie@N&G.com
County	Illinois		
Post Code	IL 4003	Type of Project	Network access
Country	USA	Date Completed	July 2000

Figure 6.14 Reference details

In Figure 6.14, we can see at a glance that the responding organization would appear to have some pertinent experience in this field and that, furthermore, they are happy to provide full details of a reference site. We shall assume that they have spoken with the reference site and confirmed that they are happy to be quoted as a reference in this context.   Discussion with the reference site can prove extremely valuable, even if problems were experienced with the original installation.   Indeed, this sort of information is extremely valuable, as the way in which a supplier organization manages perceived problems represents a good measure of both their capabilities and attitude toward business.

The next section of the response summary provides a summary of the BANTAM maps included with the response, by both map title and reference, together with any other documentation supplied with the response. Such documentation may include white papers, published product reviews, brochures or anything else that the respondent feels will be pertinent to this project. It is important to summarize such enclosures in this way, as the issuing organization may easily check that they have seen, read and taken into account, any such information thus provided.   Without a summary, they have no idea what was originally supplied.

The next section deals with estimated costs and is broken down into equipment costs, installation costs and other costs. It includes a field to note the currency expression – very important if the supplier is from another country and also a box in which other costs may be detailed. From this information, we can form a view of just how much the total implementation is likely to cost. Ordinarily, we are dealing with estimated costs at this stage, unless the issuing organization wishes to contain costs within a fixed amount and makes this clear within the RFP. We would usually expect further discussion with those successfully submitting responses to RFPs, followed by a firm offer with either fixed costs or defined rates as appropriate. This is the approach that will be taken by IFM, who will view the estimated costs as just that – a ballpark figure of how much the services being sought will cost.

The final section of the response summary contains a signatory area, where a responsible company officer can attest to the authenticity and accuracy of the proposal. It includes details of name, department, title and date signed, as well as the signature itself. Once again, it is important to capture this level of detail in order to qualify responses received and understand exactly who we are dealing with.

Having received all the responses back, IFM will hold their initial qualifying workshop and then set about analyzing the relevant proposals. There will of course be a high degree of technical evaluation, in order to understand the technical feasibility of what is being proposed, in relation to IFM's particular infrastructure. For this element of the evaluation, analysts and architects from the IT department will be consulted as necessary. In parallel with this technical evaluation should be an appraisal of the overall solution being proposed and how well it fits with the original objectives. This will be a more business or user oriented evaluation and should be undertaken very much from a usability standpoint. If both elements of the evaluation are looking good, and the estimated costs are considered reasonable, then IFM will call in the potential suppliers for further discussion. This will have the effect of producing a shortlist of perhaps just two or three potential suppliers. The next step with the IFM project, will be to arrange for each supplier to make a presentation of their proposal and its associated benefits to a carefully selected audience. Each supplier will have a finite time within which to make such a presentation and will be provided with the usual facilities to do so. When all presentations have been made, the IFM project group will set up an internal meeting, at which the successful proposal will be chosen and the respective organization invited back in order to agree the various terms and conditions under which they will work together in order to design, implement and support the application in question.

Now we are at the stage where we have conceived our solution, analyzed the market place and chosen a supplier with whom to develop the application. We can already see how useful BANTAM has been in facilitating this in an efficient and coordinated manner, both internally and between the host organization and third parties. IFM have certainly benefited from this approach and are now ready to move forward to the next stages of their project.

6.4 Understanding the systems architecture

Why should we be so concerned about systems architecture? Couldn't we just buy the solution as it is offered and implement it according to the supplier's usual way of working? Well, we could indeed do this, but there are several reasons why we should think this through more carefully. Firstly, any responsible IT department will have a defined systems architecture and associated set of principles which it follows with regard to the introduction of new applications. In addition, there will be an architectural future plan or vision which describes the way forward in this respect. Secondly, the introduction of any application which cuts across these established architectural principles, might have implications elsewhere. Thirdly, almost certainly, a new application of any significance will need to access existing data and must therefore interface in some way with existing applications or data sources. It follows, then, that any new application being considered should fit with the existing architecture. This is precisely why BANTAM includes a systems architecture map as part of its standard documentation. This allows project teams to depict the existing architecture from either a logical or actual perspective, and align this view against details of the proposed new application, in order to ensure compatibility. There are also questions of infrastructure to consider, such as network topography, communications protocols and so on. Understanding the detail of this in advance will help to ensure that the new application fits seamlessly into the existing picture.

OK, so how do we approach the question of systems architecture? The first step is to ascertain who owns the architecture and what existing documentation exists. If documentation is weak, or there is the slightest question as to its accuracy, then it should be replaced immediately. The IT support department should have a good view of this and be able to describe both the current architecture as well as the future vision. If for any reason this isn't possible, then there are some serious questions which need to be asked. More typically, it will be a question of dusting off the existing documentation and checking its currency against reality. In the case of IFM, we have already noted that they will be producing a completely new set of BANTAM systems architecture maps with which to describe the current situation, ensuring also that a process is in place for updating and revising these maps whenever a new application is added to the systems portfolio. This is a good policy which will pay dividends in ensuring that only truly compatible solutions are introduced into the organization. There is also a question of support costs here, as any incompatible or nonstandard applications would require additional levels of support and add to the complexity of the total picture.

Having understood and depicted the current architecture in a series of maps, we can then use this documentation in relation to any new application proposal. Often we will be able to use the current documents as they are. Sometimes, we may wish to produce a specific subset with which to describe a particular function or element as required for a specific project. For example, if we were working on a project for the finance department, we may wish to produce a map of all the financial systems and how they fit together. This would include any interfaces and protocols used for exchanging data, together with any links to other essential applications. From this, we could see how a new application might fit into the larger picture and how it would integrate with existing systems. We would also be able to judge any impact it may have upon existing routines.

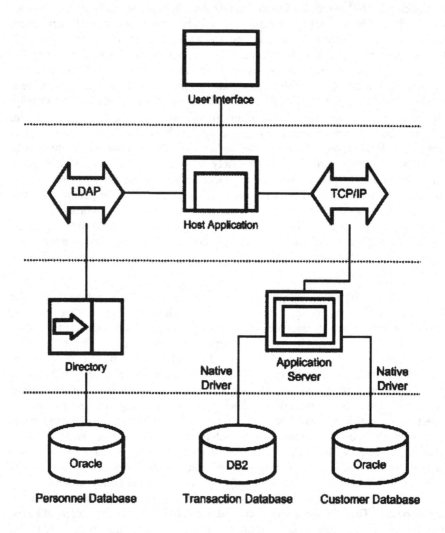

Figure 6.15 Section of a typical architecture map

In Figure 6.15, we can see an example of what a section from a typical architecture map might look like. While still at a high level, we can see that the client side application is accessing an LDAP standard directory, and that this is in turn being fed from an Oracle personnel database. We can also see that an application server is accessing data from separate transaction and customer databases via native drivers, in this case DB2 and Oracle respectively. The host application at the client side is communicating with this application server via a TCP/IP link. From this we start to see exactly how a given set of applications and data sources are working together and, therefore, how we might integrate other applications into this picture. Figure 6.15 is of course just a high level example to show the concept. In reality, you might have a much larger high level picture, supported by a series of more detailed BANTAM maps, right down to entity relationships and database schemas if required. IFM will produce a master high level map, followed by a series of smaller maps differentiated by department, which in turn will be supported by more detailed maps to show interfaces, data feeds and other such detail. All of these maps will be cross-referenced using the BANTAM reference symbol, ensuring that any relevant information may be found quickly. Remember also that each and every BANTAM map is accompanied by a Map Explanatory Notes document, providing a textual description and other information as necessary and relating to the map in question. The sum of these documents provides a comprehensive view of your organizations systems architecture which may be referenced at any time by any project team looking to implement a new application. Imagine what it would be like if you didn't have this information to hand. Every time you wanted to understand the systems infrastructure in relation to a new project initiative, you would have to undertake a complete analysis of the systems within that business area and how they fit together. This would take some time, and the chances are that you might miss an important interface or data feed if you don't have a complete picture of the situation. You would have to go through this exercise again and again with every such initiative, wasting time and resources and never really having a totally complete view. It makes a good deal of sense then, both technical and financial, to develop and maintain this master architectural view using a suitable methodology such as BANTAM.

IFM appreciate this fact, especially with regard to projects dealing with common services, such as authentication, which may cut across several application and operational areas. Furthermore, if such services use other established services, such as directories for example, then they need to integrate seamlessly together in order to provide the overall functionality required. This requires an understanding of how such a service is structured internally (its schema) and how it communicates (protocols used) with other applications. For IFM's pilot project, they intend to undertake biometric identity verification on the client device. However, they may wish to migrate this function to a central location, in wish case they will need to store the reference biometric templates in either an existing directory or a linked database, in order that they be retrieved easily in association with a user record number. Such a solution would need to be implemented in the most efficient and practical manner, taking the existing systems architecture fully into account and integrating seamlessly and securely with it. For this, you require a detailed understanding of the current systems architecture as well as the supporting infrastructure. John Getz has realized this and is happy to authorize the production of a full set of Systems Architecture Maps for IFM accordingly, knowing that these documents will be reused across many future projects, with significant potential cost savings as a result.

6.5 Getting to the detail

In the last section we looked at systems architecture and acknowledged the importance of understanding the architecture in relation to any project initiative. We also need to understand exactly how our proposed application is going to function, and therefore need to know both the logical, or operational requirements, as well as the technical detail of how the application is constructed and how it integrates with our architecture and underlying infrastructure. As we have already seen, BANTAM provides us with both Logical and Functional Scenario Maps with which to capture and document this detail.

The IFM team have already produced Logical Scenario Maps to illustrate how users will log on to the network from remote locations using the new application. They have also produced high level Functional Scenario Maps to show how this might be achieved. Now they need to provide a lower of detail, which will effectively define the application and how it is to be constructed. IFM have chosen to work in collaboration with their preferred technology supplier in order to produce this detailed understanding. This is a good approach, as it ensures that both parties are on the same wavelength and that the resulting maps will be well considered and in line with the overall requirement.

As IFM have decided to use a virtual private network (VPN) model for the communication link, one of the first steps is to understand exactly what software is needed by the client and how it will interact with the chosen operating system and the custom application software being provided for secure personal authentication. The client devices will have Windows 2000 as standard, and this operating system does have support for virtual private networks using standard protocols and the associated facilities in Windows 2000 Server. However, IFM have decided to use a dedicated hardware VPN device at the central location, with proprietary software on the client. This software will be further customized in order to integrate with the local biometric verification, which is itself integrated into the Windows 2000 security model. This will involve some cooperative work with the VPN supplier and the biometric software supplier in order to ensure everything works as planned and that the application is reliable and scaleable. It will also require close liaison with IFM's network support staff who are currently maintaining the network and are responsible for network security. The proposed application must obviously not compromise existing network security in any manner.

In order to move forward to the next stage of system definition, IFM need to produce a proposal which can act as a reference for further discussion and elaboration between the various teams involved. BANTAM can help them achieve this with Functional Scenario Maps covering the pertinent parts of the proposed system design. These maps may then go through a process of refinement and iteration, before finally being signed off as the working documents with which to design the application.

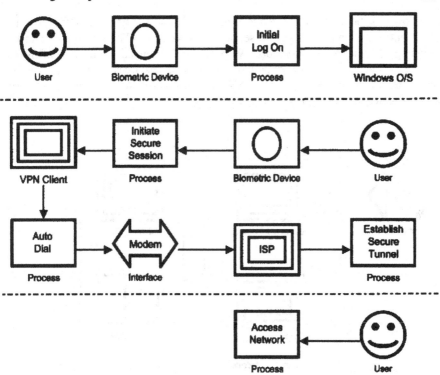

Figure 6.16 Initiating the connection

Figure 6.16 illustrates the process of initiating a connection from a remote location to the host network. In this scenario, the user logs on to the client PC using a biometric in lieu of a password. The biometric verification software resident on the client device integrates closely with the Windows 2000 security model, presenting the user with a modified log on dialog which prompts the user to supply a biometric sample at the appropriate point. If the biometric is verified against the claimed identity/user profile, then access is granted according to the permissions set for that user. If the user is directly connected to the corporate LAN at that point, then they will have access to all the network drives and resources normally assigned to them, thus functioning in a single sign on mode (assuming no further application specific access rights are required). If the client device is acting in a stand-alone, off-line mode, then the user will be able to use the applications and utilities available from the hard drive of the client device. If they wish to dial in to the corporate network from a remote location, then they need to initiate this connection from a custom dialog provided by the VPN software resident on the client. This VPN dialog in turn calls the system log on routine and requires the user to reconfirm their identity by providing a biometric sample. If the biometric is verified against the claimed identity, then a procedure is initiated to dial the ISP. The user may select the correct local number from an ISP directory in the form of a drop down list box. When the connection is made and verified, the VPN software takes control and initiates a secure session with the host VPN server via standard tunneling and encryption protocols. When the communication link is confirmed, the user has access to whatever network drives and resources they would have if they were directly connected to the LAN. This is controlled by the same directory services which control locally connected access.

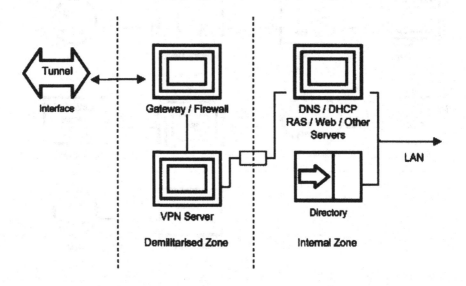

Figure 6.17 The host side connectivity

IFM prefer that the user should reaffirm their identity if they wish to connect explicitly to the corporate LAN from a remote location, especially as this connection will involve the use of an untrusted network. At the central host position, a dedicated VPN server device will sit in the demilitarized zone and will be protected by the usual firewalls and security procedures. This will be connected via routers and filters to other servers necessary to manage addresses, user permissions and other information, and finally through to the LAN, as illustrated in Figure 6.17.

In putting together this functional view of connectivity, IFM have needed to look closely at their existing infrastructure and associated inventory. Beneath the map excerpts shown in the illustrations, they have produced a more detailed architectural view showing connectivity between infrastructural components such as servers, routers and other devices, including their relative addresses and protocols used. In certain cases, it may be necessary to update or reconfigure some of these components to provide the overall infrastructure required in order to deploy the new application successfully. Discussion around the relevant BANTAM maps will drive out this sort of detail and enable the correct decisions to be made. Project Manager John Getz is happy that the work undertaken thus far using the BANTAM methodology has enabled the team to progress from a clearly articulated, but high level, original requirement through to a more detailed technical appraisal of the feasibility of implementing the proposed solution. Furthermore, this approach has facilitated a close working relationship with the technology suppliers concerned.

We have discussed just one example of 'getting to the detail', in this case looking at making a remote connection. There will of course be several other areas, both technical and process based, where the IFM team will need to develop a detailed understanding of how the system or associated process will work in practice. Much of this work may be undertaken in parallel by specialists within the project team. For example, the business analysts may look closely at the operational processes, while the network support team investigate capacity and resilience requirements. At the same time, the application development team may be looking at operating system integration and usability.

6.6 Designing the system

Having started to understand the detailed technical and operational requirements, we can turn our attention to actually designing the system. In this respect, there are several strands to be taken into consideration. Firstly, there is the software. This in itself will probably take the form of several components, including the client side software, server components, links to directory services and various interfaces. IFM have already worked out the functional requirements for these components and so will have little difficulty in developing the software from a technical perspective. However, there is the usability element to consider. In order to be well accepted among users (at least as far as it directly affects them), the software should be intuitive and straightforward to use. Once again, IFM have prepared well for this by thinking through the processes using BANTAM maps and running workshops with user representation as well as technical experts. Then there are performance and scaleability issues to consider which, again, would have been addressed when producing the BANTAM maps and aligning them against the existing infrastructure. It goes without saying that the application software should be robust and well tested on the platforms on which it will be implemented. For the IFM project, it has been agreed that the application software will be developed by the biometric supplier with close cooperation from the VPN vendor. The whole is to be overseen by IFM and built in to the overall project plan accordingly. This will naturally include a proper test routine and suitable documentation.

There is more to the system design than just the application software however. For this project, users will be issued with new notebook computers incorporating fingerprint sensors. This is a special purchase by IFM and represents a new product version from one of the leading notebook manufacturers. IFM are, in essence acting as a test site for this device and the pilot project will be monitored closely by all concerned. In the unlikely event of hardware difficulties with the integral biometric sensor, it has been agreed that a compatible external peripheral device will simply be connected to the notebook PCs and the client software configured accordingly. It is not just a question of hardware though. The notebook computers must operate as a holistic device, with seamless integration between the operating system, authentication software, communications software and the specialist operational systems used by IFM. This means providing a standard, well tested software build that can be installed and 'locked down' on each machine. The reason for fixing this standard specification is, of course, that we do not want users to alter system configuration settings inadvertently, or any parameter which might affect the secure operation of the remote client. Taking this standard build approach means that the proposed specification must be thoroughly tested against every eventuality that a user may encounter in day-to-day operation. In addition, having tested the standard build successfully, it must be properly documented for both users and ongoing technical support, or help desk personnel. This in some respects is a project in itself. The special purchase notebook PCs must also be tested for basic operational stability and suitability for use with the standard build software. All of these elements come under the broad banner of systems design and must be properly considered. For the IFM pilot project, such requirements are built into the project plan and tracked accordingly.

An associated element of systems design, which may be a little less obvious, is that of training and the provision of suitable training material. Remember, that IFM have a large workforce distributed over a diverse geography. We shall have to consider how to ensure that every employee has access, one way or another, to training for the new system. This is especially important with regard to enrolling a biometric and subsequently providing live samples for authentication purposes, not to mention the general operation of the communications software. IFM must consider both the various layers of training content, plus the media and distribution requirements. John Getz has decided that the training should be split into various layers, for technical support staff, systems administrators, local managers and users. In addition, a special 'train the trainers' program will be required. Distinct training schedules will be developed, with accreditation assigned to each one accordingly. Thus, individuals may undertake more than one training schedule where appropriate. The BANTAM Program Manager software provides a useful and intuitive way of organizing the assignment of individuals to training schedules and will be used extensively by the IFM project team in order to keep track of who has been trained and at what level. There is also a time and synchronization element to training, in that specific groups of individuals need to be trained for specific operations at the right time. This often starts with specifying and designing the training packages and then training the trainers in specific areas. From here, the different groups may be trained at the appropriate time according to project roll-out schedules. It is not useful to train people too far in advance, as they will forget much of the training if they don't have a chance to put it into practice. Similarly, we don't want to leave things to the last minute as there may be key people who can't attend, or questions that remain unanswered as the system goes live and people are required to use it in earnest. The overall project plan should therefore pay close attention to training and ensure that it is properly represented as a key part of the overall program.

Another important consideration is that of ongoing maintenance. We must try to pre-empt any problems encountered in the day-to-day operation of this system and have robust support and maintenance responses in place. Part of this may be in the form of a carefully considered help desk facility (both on line and actual) and part may be in documented maintenance procedures, including troubleshooting and component replacement where necessary. In addition, we must consider the resource requirements around delivering a robust support and maintenance facility. How many individuals do we need? What tools and training do they need? How do we support them in the field? Who do they report to? Ongoing support and maintenance is an important area and should be considered upfront as an integral part of the overall program. The support and maintenance function will, of course, make extensive use of BANTAM in documenting the processes and technical detail around the common support and maintenance issues.

In conclusion, we should think of systems design in the broader context. This is much more than just the design of the application software, important though that is. We should be thinking along two broad channels: firstly, the end-to-end solution, from software to hardware and everything in-between, including network components, both directly under our control or otherwise; secondly, what it takes to deliver this solution successfully, including training, negotiations with third party infrastructure suppliers and robust project management. All of these elements need to be 'designed' and agreed. Thus, once we have conducted the necessary analysis and research, produced our BANTAM documentation and agreed the application fundamentals, designing the system should encompass everything it takes to deliver and maintain the proposed application.

6.7 Managing the project

There are of course established methods of project management in the broader sense. There are also well established software tools that have proved popular for conventional projects. Such established methods and associated tools will coexist comfortably with BANTAM. Indeed, the use of BANTAM will complement existing methodologies by providing additional and pertinent detail to the broader picture. If your organization likes to use a particular project management methodology (maybe to tie in with resource management for example), then there is no reason why you should not append the benefits and functionality of BANTAM to an existing project management model. If you do not use a specific project management methodology, then you will find BANTAM even more useful, as it will provide the necessary discipline and clarity with which to design and run your project. In such a scenario, you will find the BANTAM Program Manager especially useful as a means of coordinating documentation and resources. You are encouraged to run the demonstration program included on the accompanying CD-ROM in order to gauge its usefulness to your organization accordingly.

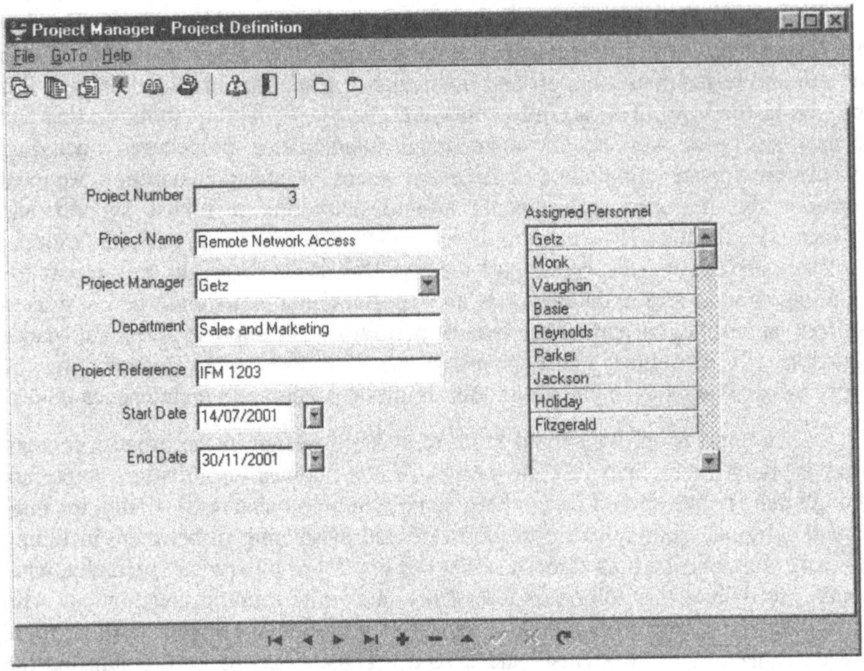

Figure 6.18 BANTAM Program Manager project definition

But what exactly do we mean by project management? Well, we could just fumble our way through the process of introducing a new system or a new way of working (or both) to our organization, addressing issues and problems as they arise and trying to find suitable personnel to deal with them. However, such an approach would be inefficient at best and may be disastrous when we are dealing with technology that has a potential impact upon existing operations. Clearly it is much better to manage such a situation carefully, with clearly defined responsibilities and standard ways of documenting and addressing issues. If follows that we also need the right people to manage the project correctly, and this starts with the project manager. The individual designated as project manager must obviously have relevant skills and experience of managing projects within the organization. Furthermore, they must be in sympathy with the objectives of the project at hand and have at least a basic understanding of the technologies and processes involved. IFM have chosen John Getz as project manager for their biometric network access project, not just for his proven project management skills and interest in the technology, but because he knows the implementation business area very well, and understands the attendant processes and issues. This allows him to take a balanced but qualified view of issues as they arise and to maintain a broad perspective. John Getz is also acutely aware of the importance of having the right people working on the project and he has taken pains to ensure a balanced team, including both business and systems analysts, programmers, product managers and business area managers who understand the current processes. To complement this core team, he will enlist the services of external suppliers or consultants as necessary in order to design and implement the application. An important element is managing these internal and external personnel efficiently and understanding exactly who is doing what and when. In addition, the interfaces and interdependencies among these different elements of the project must be understood and fully taken into account.

IFM intend to implement the first phase of this overall program in order to prove the concept, after which, a number of regional projects will be undertaken in order that the entire organization benefits from this way of working. John Getz is therefore aware that he will ultimately be managing a group of related projects and will need to coordinate both personnel and documentation across these individual projects. BANTAM is extremely useful in this respect, as lessons learned on one project may be easily documented and passed on to the next, thus steadily building experience and expertise as the overall program develops. There is a strong element of reuse here, as fundamental concepts and related technical configuration do not have to be designed from scratch each time. BANTAM maps produced for the first phase may form the basis of map documents for subsequent phases, with local editing and refinement as required. This concept of reuse would be extremely valuable for any organization with multiple branches or with a stream of related projects where certain fundamental technical concepts are common between them. BANTAM is an ideal methodology in such a case, with its clear and purposeful documentation and the ability to separate out layers of complexity.

In addition to managing resources, the project manager will need to configure a master top level project plan in order to plot the various activities against time and assign responsibilities accordingly. This will entail the creation of a set of milestones with which to define the major phases of the project. Each of these milestones will have a target date for completion and will be tracked carefully in order to manage this expectation. This is important, as there will inevitably be dependencies between the various milestones, any one of which might seriously affect overall project progress if allowed to slip significantly. John Getz has produced a simple but comprehensive spreadsheet in order to show these milestones and the individuals responsible for managing and delivering them. Many of these milestones have a direct correlation with BANTAM maps, as a completed final map document will effectively describe and define either a process or block of system functionality. This is very convenient, as the project manager can use the completed BANTAM map document to acknowledge and, if need be, confirm that the milestone has been met. This is another example of the power of BANTAM in providing a detailed audit trail, describing not only when milestones were met, but exactly how they were met. As the overall program develops through its various phases, John Getz is also aware that document management will become important. He will need to understand which documents have been used for which individual project or project phase, who created them, and of course he will need to be able to access them easily when required for meetings and workshops. For this reason, IFM have decided to use the BANTAM Program Manager as a simple but effective means of coordinating BANTAM maps, personnel, suppliers and other aspects of the program, including fundamental reporting requirements.

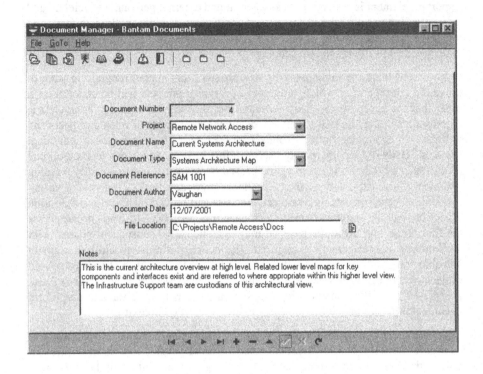

Figure 6.19 BANTAM Program Manager document management

This simple-to-use tool also provides similar facilities for assigning purchase orders and invoices to specific projects, thus creating a simple cost tracking facility. The standard reports available from within BANTAM Program Manager will also assist John Getz when he is producing his own monthly program update for the board of IFM. This is all about collating and managing information, which, for a single self-contained project, is important, but which becomes especially important for a series of projects within an overall program, or for a large project with multiple phases. The BANTAM methodology and associated tools are designed to help in exactly these sorts of scenarios. We have already mentioned the necessity for John Getz to manage the relationships between both internal and external personnel, and he will in fact produce a BANTAM Miscellaneous Definition Map in order to illustrate exactly who is doing what within this framework. He must also manage the training requirements by understanding the different elements of training required for the various groups of individuals including developers, analysts, administrators and, of course, users. In fact, training requires its own little project plan in order to synchronize with the master plan and ensure that everyone who needs to receives appropriate training at the appropriate time. The BANTAM Program Manager will be used to identify the different training schedules and log them in one place.

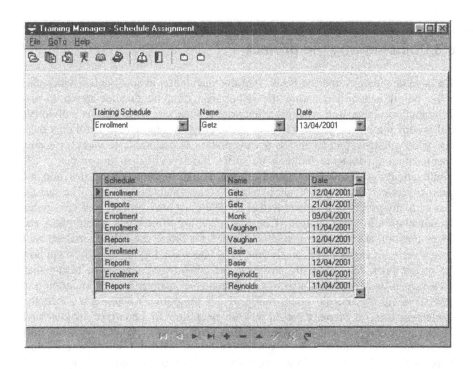

Figure 6.20 Assigning individuals to training schedules

BANTAM training schedule documents will be used to configure and describe the various schedules and the modules within them. These documents will be kept in the central repository, where they may be accessed by the project team, including a training manager who will be especially appointed in order to coordinate and manage all training related activities. This is a key position and will require an individual with a good grasp of the overall program objectives and technology, as well as proven expertise in the training field. This individual will also liaise with the analysts and consultants in order to understand the detailed requirements and gather relevant user feedback which, in turn, will be used to refine the training schedules until they are ready for final sign off and publication.

In addition to the specific areas mentioned within this section, there is obviously a requirement for good program governance throughout. This is where conventional program and project management skills come in, in order to manage risks, dependencies and overall progress. IFM are fortunate in this respect, in that they have an experienced and capable team who, together, have many years of proven experience in this area. Having such a team is important to the success of any project and is especially important when we are considering the implementation of unfamiliar technology and related processes. Strong program and project management is a prerequisite of any biometric or related initiative.

6.8 Documenting the project

We have already covered the BANTAM documentation in some detail within this book, but it is worth restating the importance of good documentation to any program. Documentation provides a blueprint against which the various individuals and groups may work in order to deliver the program objectives. It follows then, that good quality, comprehensive documentation is preferable to weak or scanty documentation. Imagine building an office block without a proper plan. We would quickly get into a mess with the detail of both fundamental construction and internal services, leading no doubt to a highly questionable end result. The same is true with any program or project initiative. We must have proper, well considered documentation to work against. Consider, also, an office block where plans and documentation are not coordinated or retained for future use. When maintenance or new services are required, much time would be wasted trying to understand how the office block was originally designed and constructed, leading to higher than necessary costs at best, and possibly poor quality of maintenance or integration of new services. Again, the same is true of our program. Assuming a successful implementation, at some stage it will be necessary to undertake maintenance checks and perhaps to enhance the system in some manner. If we have clear, high quality documentation for the maintenance teams, then this will run smoothly. This also applies to training materials and related documentation for the maintenance teams. Thinking this through at the outset and ensuring that high quality documentation is provided across the board, and delivered at the right time, will be a huge benefit to any program, and will provide considerable cost savings. This is the whole idea behind BANTAM: to provide a methodology and practical mechanism with which to document every aspect of your program initiative, from concept through to issuing RFPs and everything in between. However, there is more to documentation than simply filling in forms. In essence, we need a methodology to use the methodology. This is all about having the right attitude towards program management and documentation, and deciding right up front, exactly how you are going to document the program and how you are going to use and manage this documentation. The standard BANTAM documentation and associated notation provides a solid framework within which you can work to good effect. By choosing to use BANTAM you have already made a positive step towards good program management for your biometric or related technology program. Having done so, we must now consider the practical considerations around using such a methodology.

The first step is to decide who will be coordinating and managing the documentation on a day-to-day basis. For this, John Getz has decided to set up a small program support office who will ensure that all documentation is held within a central repository and who will also offer advice on using BANTAM to members of the individual project teams who are unfamiliar with the concept. This program support office will also be responsible for issuing all BANTAM related materials and associated software tools to those who need them. For IFM, this program support office will be run by Lucy Reynolds, who will manage all documentation and program archives. Lucy has decided to use the BANTAM Program Manager software for high level coordination and tracking of all documentation. In addition, she will make good use of the integral Personnel Manager and Supplier Manager databases in order to understand who is doing what and where they are based. This will be particularly valuable as the overall program matures and more individual projects are initiated. Lucy will be keeping a close watch on program and project milestones and reminding those responsible when related documents are due. She will also manage version control of the BANTAM documents, ensuring that each document matures into a final version and is signed off accordingly. This will be important for the program archive as well as day-to-day program operation. Finally, she will be responsible for setting up and running whatever regular reports are considered necessary for overall program management.

Reporting is an important area which is worthy of consideration right at the start, before you get involved in actually running the program. You must decide exactly how you are going to track progress and what information you need in order to do this. It is as well to be concise in our thinking in this respect. I know we live in the 'information age', but we do not want to drown in a sea of unnecessary information. Let's be clear about what we need and ensure that this is configured right from the start and that it remains consistent in both format and content throughout the life of the program. From the resource management perspective, you probably already use a system for recording individual time against projects in order to track costs. This system will undoubtedly have the ability to produce reports accordingly. However, there may be other, more program specific reports which would be useful in the context of overall program management. Such reports may include details of projects within a broader program, documents associated with a specific project, available training schedules and individuals assigned to them, purchase orders issued against specific projects and invoices received accordingly, and other such information. The BANTAM Program Manager has an integral, easy-to-use report generator to cover some of these high level requirements from a central point.

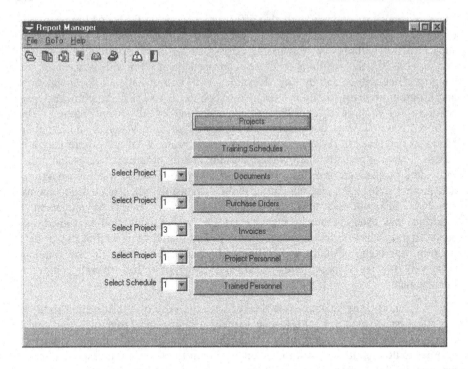

Figure 6.21 The BANTAM Report Manager

The nice thing about the BANTAM Report Manager from Lucy Reynolds' perspective, is the ease and speed with which standard reports may be created. A single button click takes care of dragging the information from the database, formatting the report and presenting it on-screen ready for printing if required. This means that in addition to producing regular monthly reports for John Getz, Lucy may easily run a report at any time if and as required for meetings *etc*. It also means that John Getz may easily keep an eye on high level external expenditure, without having to wait for reports from the finance department.

In addition to maintaining the central document repository, the program support office will design and maintain a program archive. After the successful completion of each key phase, a copy of all related documentation will be produced for archive purposes, collated and labeled accordingly. This archive may be in electronic form, paper form or both. IFM have decided that in their case it will be in electronic form, with a backup copy stored off-site. A special program archive section will also be set up on their organizational intranet, for selective access by suitably qualified individuals, via an agreed access control mechanism. IFM may implement biometric authentication for this purpose.

Some may consider that taking such pains with documentation represents an additional overhead which they don't need. The more experienced will appreciate that taking such pains actually represents a solid investment, almost certainly leading to significant cost savings in the medium and longer term. Furthermore, if your organization works to an accepted total quality standard, then the provision and maintenance of such documentation will be necessary for auditing purposes. BANTAM would therefore appear to kill several birds with one stone, providing tangible benefits at every stage of your program and continuing to provide benefits even after implementation. We have already mentioned subsequent support and maintenance for which, once again, the BANTAM documentation will prove invaluable, as it will for ongoing training purposes. In short, high quality documentation is a fundamental requirement for any successful program initiative.

6.9 Using the documentation

Having decided upon a documentation methodology, we must also understand exactly how we are going to use the resulting documents in a practical manner. The use of BANTAM facilitates this understanding due to its logical and progressive documentation set. Starting with the Application Logic and Systems Architecture Maps, these may be used very effectively during initial discussions and workshops in order to shape the high level requirements and test them against the existing infrastructure for feasibility. It is likely that these documents will be used extensively when liaising with external suppliers and consultants, as well as internally for early program or project review meetings. Furthermore, using BANTAM documents for discussion with external suppliers, provides them with the opportunity to use the same methodology in their response to any initial questions or ideas. Thus, the final versions of Application Logic Maps may be an amalgam of ideas jointly discussed and agreed between the various parties concerned, and expressed in a commonly understood format. We thus have a single, qualified high level view (the final BANTAM Application Logic Map) from which to develop lower levels of detail.

When we progress to developing the Logical and Functional Scenario Maps, a similar situation exists in that they may form the basis for discussion among the various parties involved, being incrementally refined until we are confident that we have captured the necessary detail and accuracy with which to move forward. In the case of the IFM program, they will use Logical Scenario Maps extensively to test their thinking within the various user business areas. The graphical representation plus textual description of the BANTAM maps makes it easy for nontechnical users to grasp the ideas being presented quickly, and to be able to respond in like manner, leading again to incremental refinement of proposed processes. Similarly, the Functional Scenario Maps facilitate the incremental development of the functional system components and interfaces and will be used by all those involved in this activity. The Object Association Maps may be used for entity relationship diagrams in order to aid software developers, or indeed, may be used to illustrate any relationships between objects, either physical or logical. Once again, these Object Association Maps will prove extremely useful in discussion groups and workshops as we move towards the final versions of the documents. All of the primary BANTAM maps are working documents to be actively used in the development of both ideas and actual applications design. Furthermore, they are living documents which will in themselves progress through various stages prior to maturing into the final documents, which together form the blueprint for the program. We should thus use them accordingly as the focus for our discussions and thinking throughout the design and implementation of the various phases or projects within our overall program. No doubt many BANTAM map documents will find themselves adorned with handwritten appendages and sketches as they make their way through meetings and workshops on the way towards final versions, and this is fine, as it will be achieving much in the incremental refinement of the original ideas. We may also benefit by encouraging others to use BANTAM when they are dealing with us in relation to our program, as this will help to quickly develop a common understanding of associated issues, together with a common way of expressing ideas or concerns.

The RFI and RFP documents of course serve a different purpose, although they too may be described truthfully as working and living documents. In both cases, they will form the basis for much ongoing discussion and, indeed, will be created by both the issuing and responding organizations working in tandem. They should be treated as tools within the BANTAM tool kit, used as and where applicable to help construct and deliver the program objectives. They additionally help to provide a detailed program archive, showing exactly which products and services were considered and why, which supplier organizations were chosen and why, together with all the related detail leading to such decisions.

The BANTAM training schedules are very much working documents as, once configured, they will be used extensively for training the various groups of individuals involved in the program, either as administrators or users. Final versions of these documents can only be produced with a clear and comprehensive understanding of application design and operation, together with attendant processes. They will therefore typically be subject to much discussion and development in their own right, before being finalized and published. Version control will be important here, as the training schedules and modules contained within them may be subject to change over time, as either applications or processes are refined or enhanced. This will be particularly important within a large organization, or when two or more organizations are sharing services.

We can see from the examples given throughout this book that BANTAM is an extremely practical methodology. While undoubtedly providing substantial benefits from the broader program management and documentation perspective, it is within the practical, day-to-day project related activities that BANTAM really shines and helps all concerned to move forward in a unified manner with clear understanding of application functions, operational processes and all the detail they require to deliver the original objectives. In this context, the toolbox analogy is a good one. BANTAM provides you with a very comprehensive set of tools with which to manage your program. With a little practice, these tools will enable you to be very productive indeed and produce high quality results, just as the experienced craftsman uses tools to produce superlative creations in his chosen field. As the carpenter uses his chisels and drills to hone a wooden object, you may use BANTAM to hone your thinking and application design. As the blacksmith hammers out a rough form into a finely contoured object, you may hammer out a rough concept into an efficient operational design. Use the BANTAM tools in this spirit: not for their own sake, but as a means to create a beautiful end result, in this case, a successfully implemented program which delivers positive benefits to its users and is welcomed accordingly. Be willing also to share these tools among interested parties, ensuring that you can all work effectively together towards the final creation. As BANTAM is freely available, this is in everyone's interest and can only lead towards a wider understanding of the issues, both technical and practical, around implementing biometric and related technology applications.

There are a couple of other points worth making about using the BANTAM documentation. Firstly, please use it as originally designed and provided. Do not modify any of the BANTAM document templates in any way and use only the standard BANTAM notation when creating map diagrams. This ensures that the resulting documents may be easily and immediately understood by everyone familiar with the methodology. If you have any interesting ideas for enhancing the underlying methodology or documentation, you may feed these back to the BANTAM user group where they will be given due consideration. The second point is that, as with any method, practice makes perfect. Ensure that you use BANTAM on all relative projects and you will quickly start to realize the associated benefits. High among these benefits is the introduction of a way of working which promotes clarity, consistency and reuse. The more this is embedded as a culture, the greater the benefits will become. In short, the BANTAM documents should become a familiar sight within your organization and part of your fundamental way of working with program initiatives of this kind.

7. Wrapping it up

We have covered much ground in introducing BANTAM and exploring ways in which it may be used within a typical program related situation. Naturally, every organization will have its own way of working, but within this framework BANTAM may offer substantial benefits when dealing with biometric or related technology. In fact, creative users will find that BANTAM may be used to good effect within a variety of technology related program situations.

Much of the benefit of using BANTAM may be directly associated with cost savings. If, within any project, there are misunderstandings or inconsistencies due to lack of definition or adequate communication, this usually leads to a great deal of wasted time. Time is directly related to cost, and in the context of a large program of work, such escalating costs can be considerable, sometimes bringing the program to its knees, or even killing it completely. In many instances, the original estimated cost of delivering a particular initiative bears little relation to the final cost, due in large part to program management inefficiencies and misunderstanding among those involved. As the original business case will be based upon estimated costs, this may make a mockery of the overall business plan associated with a particular program. BANTAM helps to avoid this sort of situation.

Let's appeal directly to the accountant in the organization. Using BANTAM saves money and positively impacts your bottom line. In fact, in certain cases it may help to save very considerable amounts of money. Furthermore, the basic BANTAM methodology is freely available. That sounds like a reasonable deal, but what's the catch? Simply stated, there is none. Certainly you will need to invest a small amount of time to understand BANTAM and embed the methodology into your standard working practice for projects of this nature, but the concept is so straightforward and the documentation so intuitive, that this effort will be miniscule and repaid almost immediately you start using the method.

In conclusion, BANTAM provides the missing link that has dogged many potentially valuable projects based upon emerging technologies such as biometrics and advanced token technology. The missing link in question may be described simply as the tools and working methodology which provide a common language for describing every element of such a project, and that may be used by everyone involved in the delivery of the application and associated processes. This broader team may include internal project managers and analysts, external consultants, software development teams, technology suppliers, support and maintenance personnel, training professionals, system administrators, system architects, network support personnel and others involved in the program. They can all use BANTAM and benefit from the clarity and consistency that this provides. Even the end users of the application benefit in a round about way from BANTAM, via the provision of well considered training and communication material, itself configured with reference to the original program documentation. Indeed, it is hard to find reasons not to use BANTAM within the context for which it has been designed.

7.1 The portability of BANTAM

When we refer to portability, we use the term broadly: portability within a given project, or across an entire program of related projects; portability between internal and external resources in association with a given project or program; portability across organizations and geographic boundaries; portability across related technologies.   Much of this inherent portability is provided by nature of BANTAM's relative simplicity. Anyone can understand the basic concept behind the methodology and quickly familiarize themselves with both the notation and standard document set. In addition, BANTAM is easily and freely available to all, thanks to the Internet and the distribution model this provides. This means that everyone in the chain has access to the methodology and can quickly use it to good effect with virtually no learning curve. Even the standard document templates are provided in order to get people up and running quickly.

Let's consider the implications of this for a moment. You may use the standard BANTAM documentation within your organization in order to develop the aspirations around a certain program. Your own analysts will produce map documents in order to focus discussion around processes and functionality. Your system architects will produce map documents with which to describe both the current and future technical infrastructure. Your application development teams will produce map documents in order to describe systems interfaces and dependencies. All of these documents will be interchangeable between your various internal departments and will help enormously to encourage an understanding of the bigger picture. However, the benefits don't stop there. The same documents are portable between your organization and the potential suppliers or service suppliers who need to understand the pertinent aspects of your infrastructure and application aspirations. They, in turn, may use the same documentation for internal discussion with their own technical staff, or perhaps those of a particular component manufacturer. In turn, they may produce their own BANTAM documentation which they can subsequently share with you in order to describe their proposals in a manner which is directly relevant to your particular situation and the information you have provided to them. This cross-pollination additionally facilitates the incremental development of the relative map documents, ensuring that they mature into final versions which have been carefully considered by all those concerned. Furthermore, such documents, once produced and finalized may form a valuable reference for subsequent projects which have similar operational requirements. This concept of portability and reuse is a powerful attribute of the BANTAM methodology which has the potential to provide considerable efficiencies. There is perhaps a further level of portability, and that is between similar or complementary technologies. Many of the architectural considerations particular to your organization will be the same for a variety of projects using different technology and affecting different organizational departments. BANTAM maps produced in this context may be reused for other program initiatives, providing a valuable starting point and potentially saving time. This is particularly relevant with regard to operational processes which, once properly documented, may be used by any department who needs to understand a particular process. The time that this can save within a project context can be very considerable indeed. BANTAM is thus highly portable, both as a methodology in itself, and especially with regard to the product of that methodology. When viewed in that context, its underlying value to your organization increases notably. This is of course directly relative to the amount of exposure and use that BANTAM enjoys within your organization. The more it is used, the greater the benefit realized.

Let's consider this portability from the perspective of a systems integrator or primary technology supplier. Suppose you have worked closely with a customer organization such as the fictional IFM used as an example within this book, in order to design a specific application. Having spent much time understanding the organizational infrastructure and designing a suitable system, you now have a fully documented solution for that particular application. When the next inquiry comes along for a similar requirement, you don't have to start from scratch, but can use your existing documentation as a reference against which to match the new requirement. As applications tend to fall into similar categories, such as network access, time monitoring, benefit payments, online transactions and so on, this allows you to build up systematically a library of related expertise using the BANTAM documentation. This allows you to save a considerable amount of time with each new project, which in turn allows you to be more competitive with your proposals to end users. You may wish to extrapolate this to a further level of abstraction by breaking down the BANTAM maps into single functional components, which may be reassembled in any order according to the requirements of a new application. These may be complemented by a set of typical architecture maps, depicting contemporary software and hardware combinations frequently used in the organizational sphere. With this library of proven expertise, or 'knowledge base' to use a popular term, your own organization is more properly equipped to deal with new inquiries in an effective and professional manner. A similar situation may exist for independent consultants who wish to specialize in a given area and build up their own expertise accordingly. Using BANTAM will enable them to document each hard won lesson and construct their own library accordingly. Place this on a notebook computer or PDA device and they will have a highly pertinent technical information source, instantly available wherever they may be. With the BANTAM document templates on the same device, they will also be in a position to contribute immediately to any project where the methodology is being used. There's another example of portability!

7.2 Moving forward with BANTAM

The standard BANTAM distribution, as available on the Internet and included on the accompanying CD-ROM to this book, is relatively self-contained and represents the fruits of the original BANTAM project. As the methodology becomes more widely used, it is inevitable that users will think of possible enhancements or extensions to the original version. However, it is vitally important that BANTAM is only used in its standard incarnation and not modified in any way by users. This is naturally important in order to provide the commonality and consistency which is one of the strengths of BANTAM. At the same time, we must not be blind to possible future developments, but find a way for valuable ideas in this context to be managed accordingly. This will be achieved by the provision of a BANTAM User Group (BUG). The user group will coordinate suggestions and ideas for future enhancements and ensure that these are given due consideration. It will also be a communication vehicle for details of any extensions or enhancements to the original distribution, which will be subject to strict version control. However, it is not envisaged that this would be a regular occurrence. It is also not anticipated that there would be any significant changes to BANTAM for some time, as it is necessary to reach a certain 'critical mass' of usage before feedback becomes pertinent to the wider audience. This situation will be monitored closely and we may rest assured that BANTAM will be an ongoing methodology which will evolve accordingly.

There is another aspect to moving forward with BANTAM, and that is in the context of using the methodology within your own organization. Once you have had the opportunity to use BANTAM on one or two projects, you should start to get a feel for the benefits as applicable to your particular situation and be able to shape your use of BANTAM accordingly (without deviating from the standard documentation or notation). This may take the form of inclusion within your overall program management philosophy, or distribution within the relevant organizational departments, or setting up an official document repository and archiving procedure, or some other activity. The more you use the methodology, the more refined and pertinent that usage becomes and the greater the value you will realize as a result. You may also wish to share your experiences with other organizations in order to understand and develop best practice around the use of BANTAM.

In conclusion, this book and the original BANTAM distribution, represent a first step along a road which extends beyond the immediate horizon. As the methodology becomes more widely used and appreciated for the benefits it can provide, it will undoubtedly be refined and enhanced to deliver additional functionality, as indeed required by the nature of developing technologies and how they are used within an organizational context. If you are interested in keeping abreast of such developments, then you are encouraged to join the BANTAM user group and contribute to the international BANTAM community. Details may be found within the BANTAM distribution documentation.

7.3 The accompanying CD-ROM

Included with this book is a CD-ROM containing various information and utilities which shall be described within this section. The CD-ROM file structure is as shown below.

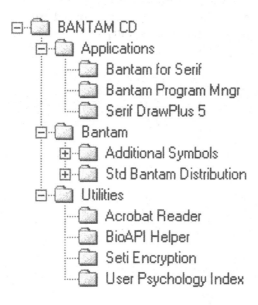

Figure 7.1 BANTAM CD-ROM file structure

The Applications folder contains trial versions of Serif DrawPlus 5 and BANTAM Program Manager. It also contains a utility named BANTAM for Serif which imports the BANTAM symbol notation into the Serif library, enabling BANTAM symbols to be dragged and dropped into the Serif DrawPlus drawing canvas. All of these applications are designed for 32 bit Microsoft Windows operating environments such as Windows 95/98/NT/2000.

Serif DrawPlus 5 is an inexpensive graphics application which nonetheless offers comprehensive functionality. While aimed predominantly at the domestic market, it is perfectly capable of creating BANTAM map diagrams, using the BANTAM symbol notation, and saving them in a variety of formats suitable for importing directly into the standard BANTAM documents. A suitable approach, for example, would be to create the map diagram using the BANTAM symbols, connected via the Serif connector tool. When the diagram is completed to your satisfaction, choose 'Edit' and then 'Select All' from the DrawPlus menu and then export the selected section into a .tif format file using the 'File' and 'Export' menu items. The resultant .tif file may be imported directly into the BANTAM documents using the 'Inset' and 'Picture' menu items from within Microsoft Word. In order to use the BANTAM symbols within DrawPlus, you must first install Serif DrawPlus 5 using the installation wizard provided. When you have installed DrawPlus and rebooted your PC, you may then run the BANTAM for Serif application. This will install the BANTAM symbols on to your PC and make them available from within Serif DrawPlus. Please use the default file locations suggested by the BANTAM for Serif installation routine. Please note that the software provided on the CD-ROM is a time limited trial version of Serif DrawPlus: see the help file for details of how to purchase a fully licensed copy.

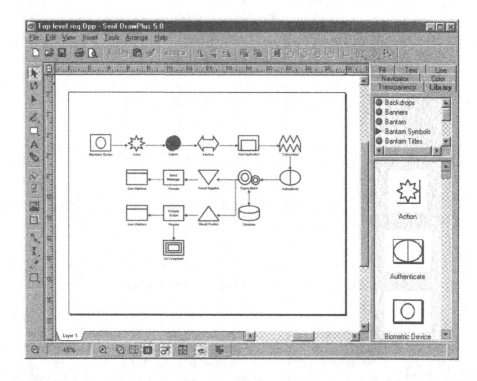

Figure 7.2 Serif DrawPlus 5

Also included in the Applications folder is a trial version of BANTAM Program Manager, an application which has been featured within various sections of this book. The BANTAM Program Manager provides a useful repository for all your program related information. It includes comprehensive personnel and supplier databases, a document management section, a project logging section, an integral report generator and much more. While offering comprehensive functionality, it remains very intuitive in use with no steep learning curve required on the part of the user. Like BANTAM itself, the software is easy to use yet deceptively powerful in the operational benefits it delivers. Use the installation wizard to install the application on to your PC and see the application itself for details of how to register and unlock the trial version.

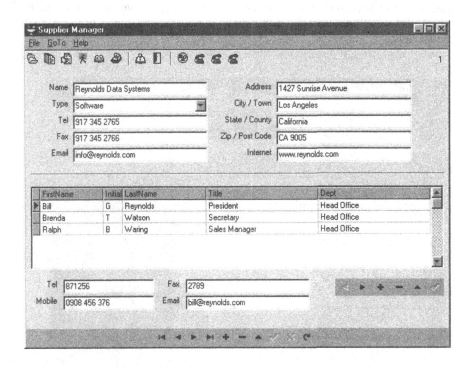

Figure 7.3 BANTAM Program Manager supplier database

The BANTAM folder on the CD-ROM contains the standard BANTAM distribution and an additional set of BANTAM symbols in Windows metafile format. The standard distribution contains a brief user guide, the notation symbols in bitmap format, a full set of document templates in both US and European page layout designs and a copy of the BANTAM Symbol Selector. The Symbol Selector is a simple utility for pasting BANTAM symbols into any Windows based application. If you already have a preferred graphics or drawing application on your PC, you may use the BANTAM Symbol Selector to paste symbols directly into that application. If you do not have a preferred application, then you may like to use Serif DrawPlus as also supplied. The additional symbol sets are provided in metafile format for use in vector based applications where scaleability might be useful. See the host applications instructions for importing metafile images. The brief guide to BANTAM is provided in Adobe portable document format (.pdf) and requires the Adobe Acrobat reader in order to view the guide. The Adobe Acrobat reader is freely available from the Adobe web site (www.adobe.com) and is also provided on the BANTAM CD-ROM for your convenience. It is recommended that you read the brief guide, even though you have read this book.

The utilities folder provides a selection of complementary applications including the Adobe Acrobat reader to which we have already referred. It also includes BioAPI helper, a useful utility for accessing BioAPI details and functions and, if required, pasting them directly into other applications or printing them. Developers will find this easier than struggling with printed copies of the BioAPI specification.

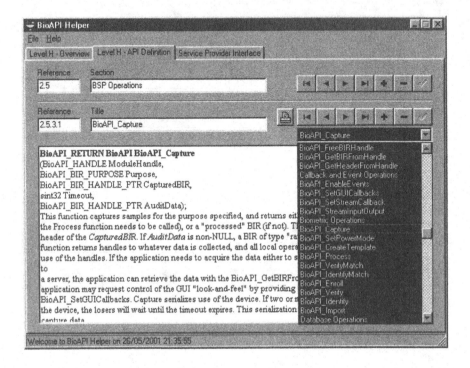

Figure 7.4 BioAPI Helper

The Seti encryption utility is a very simple application which demonstrates the principles of file encryption. It may be used to encrypt and decrypt text files using a simple password for the encryption key. You may like to experiment with the Seti application, or even use it to encrypt sensitive files or messages. Be aware however that you will not be able to decrypt encrypted files without the correct password, and be sure to read the application help file before using it to encrypt any files which are important to you.

The User Psychology Index is a very interesting application which demonstrates the effects of user psychology and environmental conditions upon the performance of biometric applications. It takes the form of an easy to use wizard that modifies the performance parameter you provide in accordance with the original User Psychology principles.

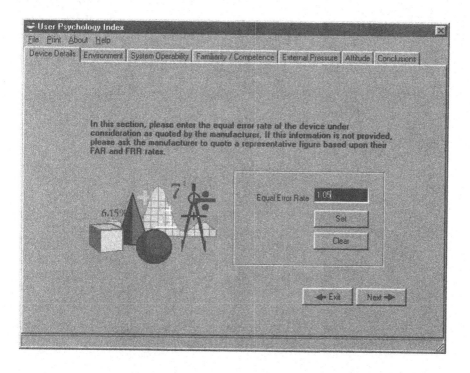

Figure 7.5 The User Psychology Index

That concludes the description of utilities provided on the accompanying CD-ROM, which hopefully you will enjoy using. Be sure to copy the standard BANTAM distribution on to your PC where it may be easily accessible, and copy the document templates into your MS Office 'templates' folder, where they may be found by Microsoft Word. If you do not use Microsoft Word, you may still be able to open these templates in your preferred word processor and them save them back as templates in the correct native format. Lastly, remember to consider joining the BANTAM User Group, as detailed in the brief user guide included with the standard distribution.

Index

Biometrics: Advanced Identity Verification

The Complete Guide
Julian Ashbourn

This is the first book to provide business readers with an easy-to-read, non-technical introduction to Biometric Identity Verification systems.

It explains the background and then tells the reader how to get their system up and running quickly. It is an invaluable read for practitioners, managers and IT personnel - in fact for anyone considering, or involved in, implementing a BIV system.

with CD-ROM

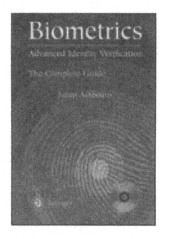

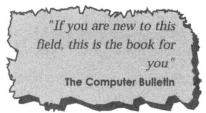

"If you are new to this field, this is the book for you"

The Computer Bulletin

All books are available, of course, from all good booksellers (who can order them even if they are not in stock), but if you have difficulties you can contact the publisher direct by telephoning +44 (0) 1483 418822 or by emailing orders@svl.co.uk

220 PAGES   SOFTCOVER WITH CD-ROM   ISBN: 1-85233-243-3